From early 1943 the General Inspekteur der Panzertruppen, Heinz Guderian, had been calling for a purpose-built tank destroyer to replace the towed anti-tank guns of the army's Panzerjäger battalions (1). Of necessity these guns operated on, or close to, the front line and casualties among the crews were consequently high. In addition, many of the weapons in use were captured Russian or French designs and in infantry formations most were still drawn by horses.

Although persuasive, there was a large degree of political manoeuvring behind Guderian's argument as a perfectly adequate vehicle had been in service with the army's independent assault gun formations since the French campaign of 1940. Designed and developed by the Berlin firm of Altmärkische Kettenwerk (Alkett) as an infantry support weapon, the Sturmgeschütz III had been successively up-armoured and up-gunned and increasingly employed as a tank killer.

But the Sturmgeschütz battalions were under the direct control of the Artillerie inspectorate which was reluctant to see any of its authority relinquished to the Inspekteur der Panzertruppen who had oversight of all Panzer, Panzergrenadier and Panzerjäger units and their arms and equipment.

Nevertheless, from mid-1943, as part of the reorganisation of the Wehrmacht's armoured units, the proposed new Panzergrenadier divisions were each allocated a Sturmgeschütz-Abteilung while the third battalions of two complete Panzer regiments were converted to Panzer-Sturmgeschütz-Abteilungen, made up of two companies of assault guns and two companies of tanks, as the first of the Gemischte, or mixed, tank battalions. At about the same time a programme was instituted to increase the firepower of the army's infantry divisions and a part of this plan entailed reinforcing each division's Panzerjäger battalion with a company of Sturmgeschütz III assault guns in an effort to offset the manpower shortages that the Wehrmacht was by then experiencing.

It soon became obvious that supply of the Sturmgeschütz III would never meet the demand to equip each division with at least ten guns and while these organisational programmes were underway, design of a fully-tracked anti-tank vehicle in the 24-ton class, initially referred to as the Sturmgeschütz neuer Art, was undertaken. This was to be a low-cost alternative, or at least supplement, to the Sturmgeschütz III (2).

The development of the new assault gun was given added impetus when the Alkett factory was bombed on 26 November 1943, severely disrupting production. Realising that the industrial centres of western Germany were now well within the range of Allied bombers, the Heereswaffenamt (HWA) investigated the possibility of producing the new assault gun at the facilities of the Böhmisch-Mährische Maschinenfabrik (BMM) plant in Czechoslovakia.

Prior to 1938 this company had been known as Ceskomoravska Kolben-Danek (CKD) and had manufactured the LT vz. 38 light tank which the Wehrmacht appropriated in large numbers as the Pzkpfw 38(t). After the German occupation the company had been absorbed by the state-owned Reichswerke Hermann Göring and renamed BMM. The Pzkpfw

Notes

1. Like many aspects of this story, Guderian's appointments are a source of contention. In his autobiography he maintains that he took up the post of General Inspekteur der Panzertruppen on 1 March 1943. He was promoted to Chef des Generalstabs des Heeres, Chief of the General Staff of the Army, on 21 July 1944 but continued in both roles until 29 March 1944 when he went on sick leave.

2. The cost of building a Sturmgeschütz III ausf G was 82,500 Reichsmarks which, adjusted for inflation, would be almost US$1,500,000 today. Unlike Alkett, the major producers of the Hetzer were largely controlled of the German government.

Photographed on 20 April 1944, the occasion of his birthday, Hitler inspects one of the first production Hetzers. This vehicle is fitted with the early Kugellaffette I ball mount but lacks the gun mantlet. Most of the first production batch were missing armoured parts or were defective in other ways and were returned to the factory. Note that the lead for the Notek driving light enters the hull through the roof, a feature that was thought to be restricted to the prototypes.

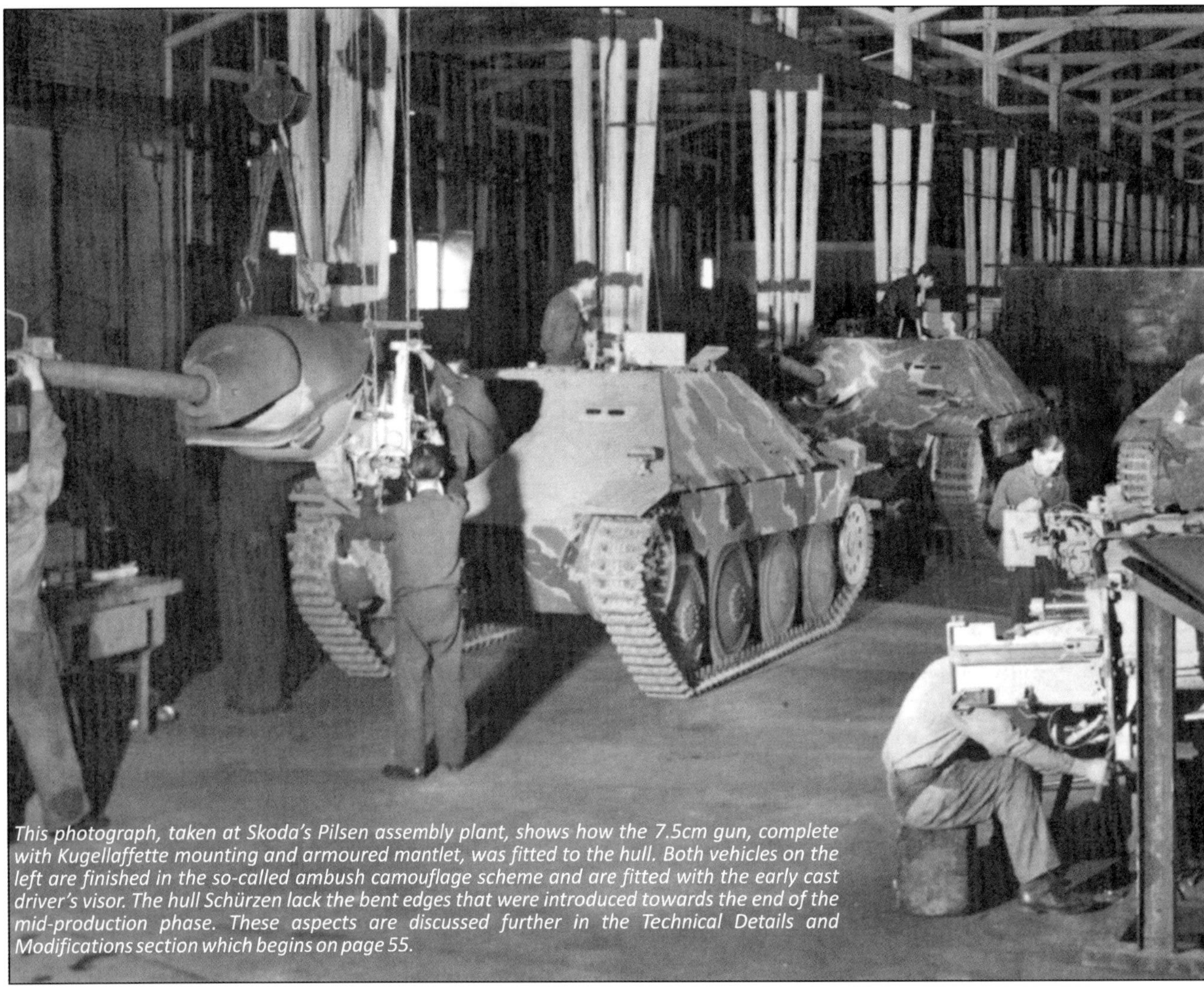

This photograph, taken at Skoda's Pilsen assembly plant, shows how the 7.5cm gun, complete with Kugellaffette mounting and armoured mantlet, was fitted to the hull. Both vehicles on the left are finished in the so-called ambush camouflage scheme and are fitted with the early cast driver's visor. The hull Schürzen lack the bent edges that were introduced towards the end of the mid-production phase. These aspects are discussed further in the Technical Details and Modifications section which begins on page 55.

38(t) had remained in production until June 1942 and although lightly armed, was well liked for its reliability and ease of maintenance. The hulls and chassis were used as the basis for a number of self-propelled gun designs from early 1942 until mid-1944.

Although Hitler was enthusiastic about the plan to utilise BMM, preliminary planning showed that the company had neither the lifting capacity nor space to build a 24-ton assault gun. At a meeting held on 6 December 1943, the Führer agreed that the weight could be no more than 13 tons, a stipulation that meant that a complete redesign with drastically reduced armour protection was required. To compensate, the armour plates would be steeply sloped and the vehicle would be capable of speeds of up to 60 kilometres per hour.

As components for both the Pzkpfw 38(t) tank and the Pzkpfw 38(t) neuer Art reconnaissance vehicle were available at BMM it was decided to utilise these in the development of the new assault gun and so the leichte Sturmgeschütz 38(t) programme was instigated. Guderian and his staff stubbornly refused to refer to the proposed design as an assault gun and in all the correspondence of the Inspekteur der Panzertruppen the phrase leichte Panzerjäger 38(t) was used. By the time the order for production had been placed it seems that the name, Panzerjäger 38 für 7.5cm Pak 39 (L/48) (Sdkfz 138/2), had been decided upon although the term leichte Sturmgeschütz 38(t) was still being employed, by Hitler at least, at the same time.

It was not until 11 September 1944 that the name Jagdpanzer 38 - Panzerjäger 38 für 7.5cm Pak 39 (L/48) (Sdkfz 138/2) was officially adopted and in the November 1944 HWA report this was shortened to Jagdpanzer 38 (m. 7.5cm Pak 39 L/48) (Sdkfz 138/2).

The name Hetzer was definitely used by operational units, as several surviving documents prove, but it was not a term coined by troops in the field as Guderian suggested to Hitler in December 1944 (1).

Rather, it was the name assigned to the E-10 project and as this vehicle may have been mentioned in the first meeting of the designers at BMM and representatives of the HWA, its later use when referring to the Jagdpanzer 38 could have been the result of a simple misunderstanding (2).

As a matter of simple convenience I have used the terms Hetzer and Jagdpanzer 38 throughout the remainder of this book.

Notes

1. Reports submitted by Panzerjäger-Abteilung 743 in July 1944 use the word Hetzer but by August the reporting officer was using the term le.Pz.Jg.38t. As late as April 1945, the Panzerjäger-Abteilung of the Waffen-SS Nibelungen Division stated that ten Jg.Pz.38t were on hand. Interestingly, it appears that the 't' suffix, denoting Tschechisch or Czech, was never entirely discarded.

2. The E-10 was envisaged as a very low silhouette tank destroyer with a top speed of 70 kilometres per hour. Its development did not progress beyond the drawing board.

As the suspension and lower hull were already available, the development of the 13-ton Sturmgeschütz went forward at an unprecedented rate. By 17 December 1943 the design drawings were complete and within a week a wooden mock-up had been presented to the representatives of the HWA and it was at that time that the decision was made to use the 7.5cm Pak 39 L/48 gun, which was also being employed in the Jagdpanzer IV programme. Oberst Wolfgang Thomale, General Guderian's deputy, ordered the construction of three prototype vehicles to be ready for troop trials by March 1944. As all the automotive components had been thoroughly tested and proven with the Pzkpfw 38(t) series it was felt that pre-production testing was unnecessary.

By 18 January 1944, BMM had been contracted to build 1,000 leichte Sturmgeschütz 38(t) and an ambitious production schedule was set that would have seen 500 vehicles per month leaving the factory's assembly lines by March 1945. But the rushed design and development process meant that the first vehicles were plagued by minor faults including leaking gaskets, constricted fuel lines and problems with the carburettor and air filtration system until at least August 1944. As part of the remedy the HWA reduced the production goals at BMM and in July contracted Skoda Werke, another Czech firm under the control of the Reichswerke Hermann Göring, to build 2,000 additional vehicles. A detailed examination of the production and allocation numbers is presented in the accompanying chart.

The production model had a very low profile at just over 2 metres, which included the shield of the Rundumsfeuer machine gun, and was relatively short at 4.766 metres. These advantages were paid for by a cramped interior and for this reason the vehicle was not popular with its crews. The normal firing height of the main gun was just 1.4 metres above the ground and this caused problems with the dust thrown up from the blast. But the 7.5cm L/48 gun had a flat trajectory of almost 1,000 metres and this was enough to kill a Soviet T-34 tank at long range with a hit on its frontal armour.

The Hetzer was constantly modified during the course of its production life, often due to economy measures, and this is examined in the Technical Details & Modifications section of this book which begins on page 55.

Although it was only ever envisaged as an interim solution, the Hetzer proved to be one of the most successful tank destroyers of the Second World War period. Introduced into the Wehrmacht at a time when defensive fighting was the norm, its low silhouette, sloped armour and 7.5cm gun allowed the diminutive Jagdpanzer to take on much heavier opponents and its designed has influenced the development of tank hunting vehicles to the present day.

JAGDPANZER 38 PRODUCTION AND ALLOCATION

Month	BMM Production *Fgst series 321001-323000*	Skoda Production *Fgst series 323001-325000*	Bergepanzer 38 Production	Accepted Heereswaffenamt	Allocated Heereszeugamt (1)
March 1944	3 (2)				
April 1944	20			23	
May 1944	50			50	22H (4)
June 1944	100			100	8H (5)
July 1944	100	10		100	119H, 4B
August 1944	150	20		171	122H, 2B
September 1944	190	30		124	158H
October 1944	133	57		290	193H, 31B
November 1944	298	89		403	290H, 27B
December 1944	223	104		327	323H, 16B (6)
January 1945	289	145		434	409H, 11B (7)
February 1945	273	125	43 (3)	398	312H, 39B (8)
March 1945	148	153	21	301	463H, 27B
April 1945	70	47	12	Unknown	151H, 12B (9)

1. The figures presented here were taken from the records of the Heereszeugamt, the German Army's supply directorate. The suffix H reresents Hetzers while B signifies a number of Bergepanzer 38 recovery vehicles. Command tanks were not shown separately. It should be noted that vehicles allocated in the last days of a particular month may not have been shipped until the following month. 2. All prototypes. 3. This is the first time the Bergepanzers are reported separately. Previous assemblies were included in the BMM total. 4. All allocated tp training facilities. 5. All allocated to the Ersatz-Heer. 6. Fifty Hetzers were shipped to the Hungarian army on 7 and 10 December 1944. This total also includes twenty Flammpanzer 38. 7. This total includes thirty-two vehicles returned from long-term repair. 8. This total includes thirty vehicles returned from long-term repair. 9. As few as 121 of these vehicles may have actually been shipped. Additional records compiled by a section of the Inspekteur der Panzertruppen show a further ninety-four Hetzers and two Bergpanzer 38 allocated at some time after 15 April 1945.

This Jagdpanzer 38, Fgst. Nr. 322111, was assembled at BMM in early December 1944 on a Skoda-manufactured hull. Note the camouflage pattern, which is described in the Technical Details & Modifications section, and the black rectangles painted onto the hull front below the driver's visor. The MG 34 machine gun and the sight for the main gun are missing and the spare tracks were not normally carried in the position seen here. This vehicle is today on display at the Bovington Tank Museum.

Notes

1. A number of the reformed units had originally been equipped with the Sturmgeschütz III assault gun and continued to be referred to Sturmgeschütz-Abteilung, Sturmgeschütz-Kompanie or even Sturmgeschütz-Batterie.
2. Truppenübungsplatz Mielau, modern-day Mlawa in Poland, was one of four large training facilities in East Prussia. It was also known as Truppenübungsplatz Nord.

From late 1943, in an effort to offset the severe manpower shortages that the Wehrmacht was by then experiencing, a programme was instituted to increase the firepower of the army's infantry divisions and a part of this plan entailed reinforcing each division's Panzerjäger battalion with a company of assault guns.

Within the battalion these companies were usually, but not always, identified by a unique number arrived at by adding 1000 to the number of the parent formation. For example, 2.Kompanie of Panzerjäger-Abteilung 167, the assault gun company, was referred to as Panzerjäger-Kompanie 1167 (1). In addition, a small number of battalions, usually attached to Volksgrenadier divisions, received two assault gun companies. Where these instances are known I have made mention of it.

Note that the list does not contain units which served for the most part in Italy or on the Eastern Front. I should also mention that a number of the units mentioned here are also discussed in TankCraft 19, *STUG III & IV: German Army, Waffen-SS and Luftwaffe Western Front, 1944-1945.*

Panzerjäger-Abteilung 3. Attached to 3. Panzergrenadier-Division, the battalion was destroyed at Stalingrad and rebuilt in January 1943 with one Flak and two Panzerjäger companies.

In at least one authoritative source this unit is confused with Panzer-Jagd-Abteilung (gp) 3 which was allocated twenty-one Jagdpanzer 38 Hetzers on 7 April 1945. Panzerjäger-Abteilung 3 operated only Marder II, Jagdpanzer IV and Panzer IV/70(V) tank destroyers and it is included here only to correct the error.

Panzerjäger-Abteilung 9. Attached to 9.Infanterie-Division, a single company of the battalion was located at Mielau, retraining as an assault gun unit, when the division was destroyed on the Eastern Front in August 1944 (2).

The company remained in Poland and on 18 November was allocated fourteen Hetzers. Renamed Panzerjäger-Kompanie 1009, the company joined the newly-raised Panzerjäger battalion of 9.Volksgrenadier-Division and took part in the Ardennes Offensive, most notably in the fighting around Nothum and Wiltz.

By March 1945 the division was worn out and the remnants had been absorbed by 352.Volksgrenadier-Division.

Panzerjäger-Abteilung 26. The battalion was destroyed in the fighting on the Eastern Front and rebuilt around 2.Kompanie, Panzerjäger-Abteilung 1582 in September 1944. In November the battalion's assault gun company, referred to as Sturmgeschütz-Abteilung 1026, reported that fourteen Hetzers were on hand.

The battalion was attached to 26.Volksgrenadier-Division which was engaged in heavy fighting during the Ardennes Offensive in December 1944 and

...........*text continued on page 6*

PANZERJÄGER-ABTEILUNG, VOLKSGRENADIER-DIVISION, 1944

From late 1943 the Oberkommando der Wehrmacht (OKW), faced with mounting personnel losses, embarked on a programme to expand the overall firepower of the army's infantry divisions. In addition to an increase in the number of automatic weapons available to infantry formations, it was intended that each division's Panzerjäger battalion be allocated a company of ten or fourteen assault guns. All formations of the German Army were organised according to detailed tables of establishment referred to as Kriegstärkenachweisung, usually abbreviated to KstN, which showed the official composition of a unit, listing the exact number of personnel and type of equipment from vehicles to small arms. The organisation shown here was specifically created for a Volksgrenadier-Division, the first of which were raised in the autumn of 1944, but the battalions of Infanterie, Grenadier and Gebirgs divisions were essentially the same. During this period it was planned that the army's independent Sturmgeschütz battalions would be converted to Heeres-Sturmartillerie-Brigaden and assault gun companies of divisional Panzerjäger battalions were now referred to as Sturmgeschütz-Abteilungen and uniquely numbered within the division by adding 1000 to the division's number. For example, the assault gun company of 559.Volksgrenadier-Division was referred to as Sturmgeschütz-Abteilung 1559. In a small number of divisions the first company was also reorganised as an assault gun company such as 212.Volksgrenadier-Division where the companies were referred to as Sturmgeschütz-Abteilungen 1212a and 1212b. These are also described in the main text. From October 1944, in the official allocation lists at least, the companies were increasingly referred to as Panzerjäger-Kompanien and this may have been applied to battalions which were assigned the Jagdpanzer 38 or Jagdpanzer IV L/48 in place of the Sturmgeschütz III assault gun. There were, of course, exceptions to these rule and they are mentioned in the unit histories. An important factor which should be mentioned is that there were never enough armoured vehicles to fulfil requirements and many divisions did not received their authorised tank destroyers, relying instead on motorised companies or even horse-drawn transport.

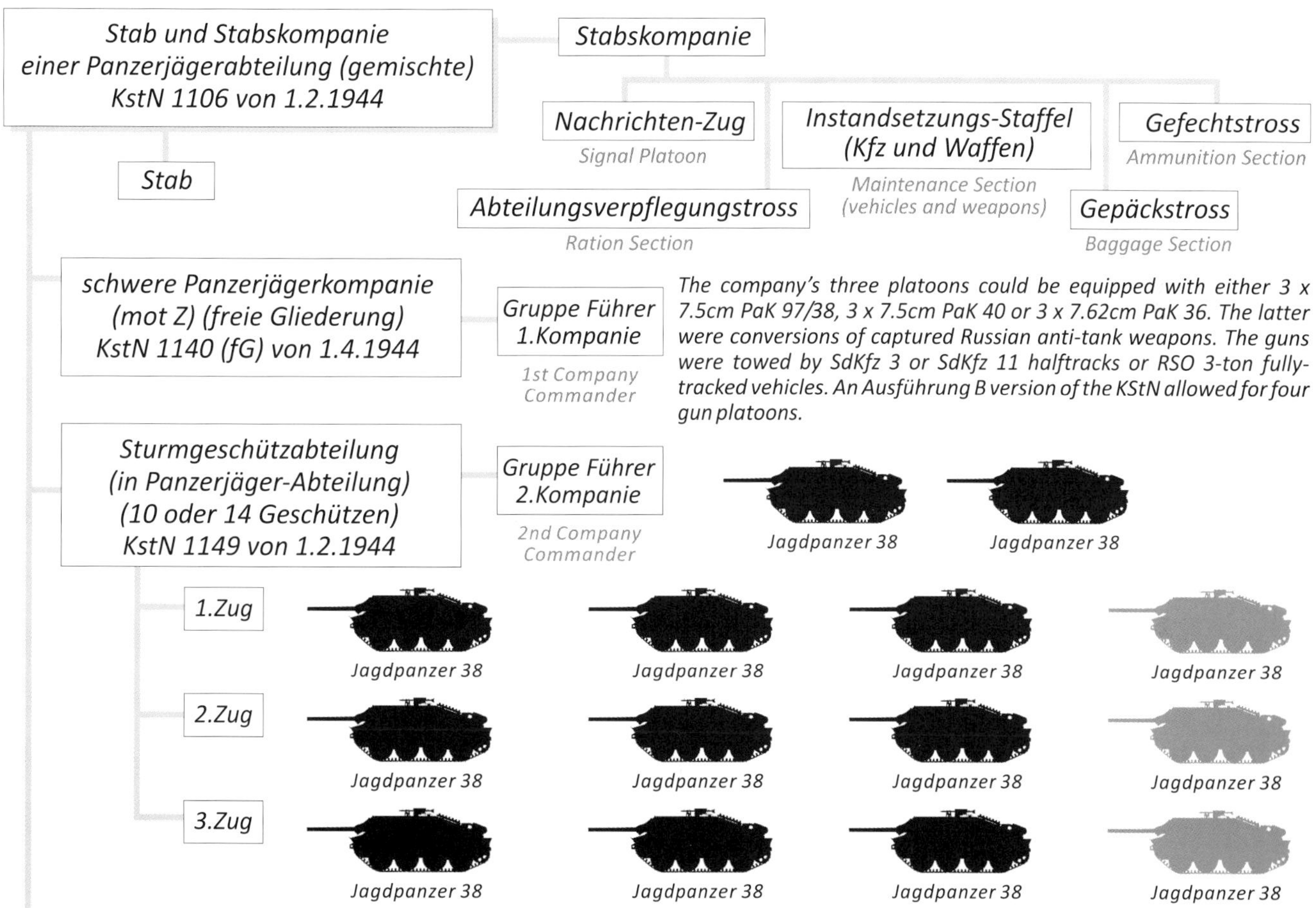

Note that although this KstN could be filled with either Sturmgeschütz III, Jagdpanzer IV L/49 or Jagdpanzer 38 tank destroyers, as shown here, the type of vehicle would be consistent throughout the company, including the headquarters. The vehicles depicted in a lighter shade would be added only for the Ausführung B version of this KstN, which allowed for four-gun platoons, giving an overall total of fourteen. In the official order the ten gun company was referred to as Ausführung A.

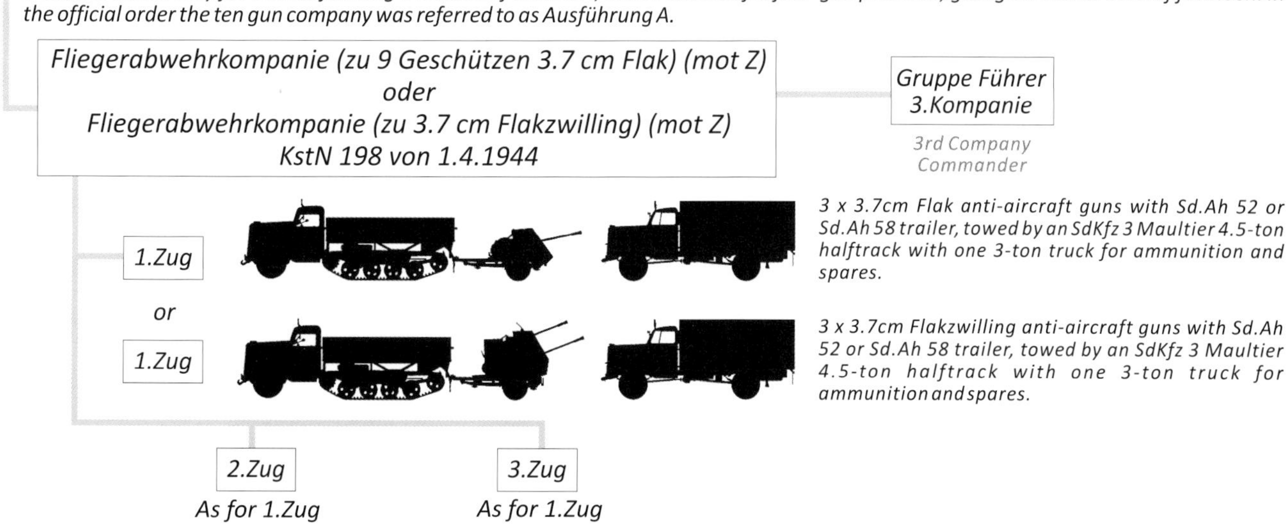

............text continued from page 4

spent much of January 1945 trying to break into the American defences around Bastogne, suffering heavy losses. Retreating into Germany, the division was in the Harz Mountain region when the war ended.

Gebirgs-Panzerjäger-Abteilung 55. Attached to 2.Gebirgs-Division, the battalion was allocated fourteen Hetzers on the last day of 1944 and these had all arrived by the end of January 1945. These were all assigned to the battalion's 2.Kompanie which was also referred to as Gebirgs-Panzerjäger-Kompanie 1055. Organisational charts, or Gliederung, produced by the division show that seven vehicles were fully operational while four were in repair in mid-February and that just four vehicles in total were on hand by the beginning of March.

Panzerjäger-Abteilung 106. See Panzerjäger-Abteilung Lambert below.

Panzerjäger-Abteilung 147. Raised in late August 1944 in Denmark as 2.Kompanie, Panzerjäger-Abteilung 1577, the unit was expanded and renamed Panzerjäger-Abteilung 147 in the following September and later attached to 47.Volksgrenadier-Division.

On 6 November the battalion was allocated fourteen Hetzers and these were reported as being on hand in the following month. The Hetzer company was referred to as Sturmgeschütz-Kompanie 1147.

Half of the division was made up from surplus Luftwaffe and Kriegsmarine personnel while about another third consisted of recently drafted 17 and 18 year-old boys. The first elements of the division arrived on the Western Front on 16 November 1944, unloading the men and equipment at Düren, west of Cologne, which was at that time defended by units of 5.Panzerarmee. That night the town was bombed and the German units suffered heavy losses including many officers and men of the division's headquarters.

Although the division was supposed to man the defensive positions along the Westwall it was fed into the fighting piecemeal and by 21 November was so badly depleted that the remnants were merged with 12.Volsgrenadier-Division to form an ad-hoc Kampfgruppe. We must assume that the Panzerjäger battalion took little part in the fighting as Sturmgeschütz-Kompanie 1147 reported that the original fourteen Hetzers were still on hand in December. Elements of the division remained in this area taking part in the last phase of Operation Nordwind.

But by the beginning of March 1945 the division's strength was less than 400 men and when French units attempted to cross the Rhine at Germersheim what remained of 47.Volsgrenadier-Division was too weak to stop them. On 24 April 1945 the Kampfgruppe managed to cross to the eastern bank of the Danube but was almost completely destroyed in the process and only isolated stragglers surrendered to the advancing Americans.

Panzerjäger-Abteilung 162. Attached to 62.Volksgrenadier-Division, the battalion had been completely destroyed in the fighting on the Eastern Front and rebuilt in September 1944.

The battalion's assault gun company, referred to as Sturmgeschütz-Kompanie 1162, was allocated fourteen Hetzers on 17 November 1944 and these were reported as being on hand in the following month.

The division was assigned to LXVI.Panzerkorps and took part in the Ardennes Offensive where it was engaged in some of the heaviest fighting, including the attempts to cross the Our River at Steinebruck and the assaults on St Vith. By mid-January 1945 the division was slowly retreating towards the German border. In March the battalion was allocated eight new Hetzers and the second company was referred to in the allocation lists as Jagdpanzer-Kompanie 1162.

By April the remnants of the division, reinforced by boys of the Hitler Jugend from Düsseldorf and Cologne, were fighting in the Ruhr Pocket battles and ended the war there.

Panzerjäger-Abteilung 167. The battalion was formed in October 1944 from elements of the anti-tank battalion of Schatten-Division Niedergörsdorf and attached to 167.Volksgrenadier-Division (1). The battalion's assault gun company was referred to Panzerjäger-Kompanie 1167.

On 25 November 1944 the Heereszeugamt (HZA) shipped fourteen Hetzers to the battalion and these were all on hand when the division took part in the Ardennes Offensive, initially held in the OKW reserve (2). Vehicles of the unit are shown on pages 18 and 19 of the Camouflage & Markings section of this book.

From 28 December the division, with Panzergrenadier-Lehr-Regiment 901 attached, supported the units of 1.SS-Panzer-Division in the attacks on Bastogne . The division performed well but suffered heavy losses, the US Army official history describing the division's initial assault as being 'cut to pieces' by artillery fire and fighter-bomber attacks.

The remnants of the division were engaged in the defensive battles along the Westwall and were trapped in the Ruhr Pocket in April 1945.

Notes

1. The Schatten, literally shadow, divisions were very loosely formed units that supplied partly-organised formations and trained replacements to front-line divisions. They usually contained infantry, artillery, engineer and anti-tank units but no staff or supply elements.
2. The Heereszeugamt was the army's supply directorate.

Panzerjäger-Abteilung 179. The battalion's assault gun company, Panzerjäger-Kompanie 1179, had in fact been one of the first units to be issued the Hetzer in August 1944 but by the time the company had completed its training its parent formation, 79.Infanterie-Division, had been was annihilated and the company operated for some time as an independent unit on the Eastern Front.

The division was later rebuilt as 79.Volksgrenadier-Division but Panzerjäger-Kompanie 1179, which had been allocated fourteen Hetzers on 7 December 1944, was at about that time subordinated to 257.Volksgrenadier-Division where it probably remained.

The assault gun company's place was filled by Panzerjäger-Kompanie 1352b, which was the second assault gun company of Panzerjäger-Abteilung 352, one of a number of anti-tank battalions raised with four companies.

Panzerjäger-Abteilung 212. Attempts to rebuilt the battalion after the fighting on the Eastern Front from survivors of the original formation and parts of Panzerjäger-Abteilung 13(L), a Luftwaffe unit, were abandoned in September 1944 and a new battalion was created from Panzerjäger-Abteilung 1578.

The battalion was attached to 212.Volksgrenadier-Division which took part in the Ardennes Offensive and the defensive battles of early 1945. This division was one of a small number of Volksgrenadier formations formed with a Panzerjäger battalion that contained two assault gun companies. Confusingly they were referred to as Sturmgeschütz-Abteilungen 1212a and 1212b and were equipped with Sturmgeschütz IV and III assault guns respectively. At least one source suggests that four Jagdpanzer 38 Hetzers were on hand in April 1945 but this is almost certainly incorrect.

Panzerjäger-Abteilung 219. The battalion was formed in September 1944 and attached to 183.Volksgrenadier-Division. The second company was referred to as Sturmgeschütz-Kompanie 1219 and was allocated its first fourteen Hetzers on 9 September and these were reported as arriving in the same month. On 14 December another four vehicles were dispatched by the HZA. The division took part in the fighting for the Siegfried Line and in the defence of Aachen. In April 1945 it was trapped in the Ruhr Pocket.

Panzerjäger-Abteilung 235. See Jagdpanzer-Kompanie 1235 below.

Panzerjäger-Abteilung 245. See Jagdpanzer-Kompanie 1245 below.

Panzerjäger-Abteilung 246. Formed from elements of Panzerjäger-Abteilung 168 in September 1944, the battalion was attached to 246.Volksgrenadier-Division. On 9 September the battalion was allocated fourteen Hetzers and these were all assigned to the second company which was referred to as Sturmgeschütz-Abteilung 1246. The division took part in the defence of Aachen but surrendered in mid-October 1944. Rebuilt in November,

...........*text continued on page 9*

Photographed outside Kilstett, north-east of Strasbourg, in January 1945, this Jagdpanzer is often associated with Panzerjäger-Abteilung 553 and although 553.Volksgrenadier-Division did take part in the fighting north of Strasbourg, as part of Operation Nordwind, its Panzerjäger battalion did not receive any of these vehicles before the end of March 1945. The OKW situation maps that have survived for this period show only the Volksgrenadier division and a Kampfgruppe made up of elements of 10.SS-Panzer-Division in the area. Also mentioned is Division Nr 905, which was little more that a divisional staff but did control a unit referred to as leichte SS-Panzerjager-Kompanie 95. It is possible, although far from certain, that this company had some connection to the Hetzers that were allocated to Heeres-Gruppe Oberrhein/Reichsführer-SS in December 1944 and January 1945.

Identifying markings on either vehicle, the only unit close to the town at that time equipped with the Hetzer, was the Sturmgeschütz company of Panzerjäger-Abteilung 1316 from 16.Volksgrenadier-Division which had originally been issued with fourteen of these tank destroyers in early December. The factory-applied camouflage is typical of vehicles manufactured by the Bömisch-Märische Maschinenfabrik (BMM) works and visible in the original print of this photograph, just to the left of the damaged Notek headlight is a matte black rectangle which, from October 1944, was painted on the hull in an effort to decoy enemy gunners away from the driver's visor and vision block.

............*text continued from page 7*

the division returned to the front and was engaged in the battles in the Hürtgen Forest. In December a total of four Hetzers were shipped to the Panzerjäger battalion.

Panzerjäger-Abteilung 257. The battalion had been attached to 257.Infanterie-Division but 2.Kompanie, referred to as Sturmgeschütz-Abteilung 1257,was training at Milowitz in Czechoslovakia when its parent formation was destroyed in the fighting in Romania in August 1944 (1).

The division was rebuilt in October but the assault gun company never returned and was replaced by Panzerjäger-Kompanie 1179 which was allocated fourteen Hetzers on 7 December 1944. See also Jagdpanzer-Kompanie 1257 below.

Panzerjäger-Abteilung 272. Formed in September 1944 from elements of 2.Kompanie, Panzerjäger-Abteilung 1575 the battalion was attached to 272.Volksgrenadier-Division. The assault gun company was referred to as Panzerjäger-Kompanie 1272.

Fourteen Hetzers were shipped from the HZA on 28 October and these were all on hand by the end of November. After just six weeks of training the division was sent to the front and took part in the battles in the Hürtgen Forest and along the Roer River before retreating to the Rhine.

For much of March 1945 the division was engaged in the battles to hold the defensive line along the Rhine River, first at Niederbreisig then at Waldorf, south of Brohl, and finally on the southern section of the Remagen bridgehead.

At this time six new Hetzers were dispatched to Panzerjäger-Kompanie 1272 directly from Nachschub OB West (2). Although this reinforcement was welcomed, by this time the division's combat strength had been reduced to 300 men and on 12 April it was officially disbanded.

Panzerjäger-Abteilung 276. Initially formed in January 1944 the battalion was completely destroyed in the Normandy battles. In September the battalion was rebuilt from elements of Panzerjäger-Abteilung 580 and attached to 276.Volksgrenadier-Division. The formation of the battalion's self-propelled company, referred to as Panzerjäger-Kompanie 1276, was cancelled on 24 November 1944 and the Jagdpanzer 38 Hetzers that were earmarked for the company were sent elsewhere (3).

Panzerjäger-Abteilung 277. Formed in September 1944, the battalion was attached to 277. Volksgrenadier-Division. The battalion's second company was raised from 2.Kompanie, Panzerjäger-Abteilung 1574 and was referred to as Sturmgeschütz-Abteilung 1277. On 23 October 1944 the battalion was allocated fourteen Hetzers and these had all arrived by the end of November. The division took part in the Ardennes Offensive and was tasked with clearing the Hollerath to Losheim road as far as Elsenborn for Kampfgruppe Müller of 12. SS-Panzer-Division.

The fighting here was so costly that the division was delegated secondary roles for the remainder of the battle and was eventually relieved by units of 3. Panzergrenadier-Division. In January 1945 the division was pushed back to its starting positions and by February the infantry strength had been reduced to just 800 men. The remnants of the division ended the war fighting in the Ruhr Pocket battles in April 1945.

Panzerjäger-Abteilung 326. Originally formed in January 1944, the battalion suffered heavily in the Normandy fighting and the survivors were incorporated into Artillerie-Regiment 1579. Reformed in September 1944, the battalion was attached to 326.Volksgrenadier-Division. The battalion's second company, referred to as Panzerjäger-Kompanie 1326, was allocated fourteen Hetzers on 19 November and the vehicles probably arrived later in the month (4).

The division was sent to the Western Front in December 1944 and fought in the battles along the Roer River and took part in the Ardennes Offensive. The surviving elements surrendered in the Ruhr pocket in April 1945.

Panzerjäger-Abteilung 340. Attached to 340.Volksgrenadier-Division, the battalion was formed in September 1944. The battalion's second company, which was referred to as Panzerjäger-Kompanie 1340, was allocated fourteen Hetzers by the HZA on 15 November 1944 and these were reported as being on hand in December. The division took part in the fighting around Aachen and later in the Ardennes where it was subordinated to I.SS-Panzerkorps.

The division fought in the defensive battles of early 1945 and, as part of XIII.SS-Armeekorps, was destroyed in the Ruhr Pocket in April.

Panzerjäger-Abteilung 352. Originally formed in November 1943, the battalion was completely wiped out in the fighting in Normandy and rebuilt in September 1944 with four companies from the remnants of Panzerjäger-Abteilung 1581. Fourteen Jagdpanzer 38 Hetzers were shipped to the battalion on 13 November and these were all assigned to Panzerjäger-Kompanie 1352a. A further fourteen vehicles were allocated by the HZA to

Notes

1. Milowitz, modern-day Milovice in the Czech Republic, was the site of a massive training facility that had been taken over by the Wehrmacht in 1939. It housed a number of armour instructional units including Panzerjäger-Kompanie Milowitz and Panzerjäger-Einheit H der Panzerjägerschule Milowitz.

2. Nachschüben were supply depots. The particular example mentioned here was under the direction of Oberbefehlshaber (OB) West, the overall military commander on the Western Front.

3. Confusingly, three Panzerjäger companies were raised with the number 1276: the unit mentioned here, 1276a, equipped with Jagdpanzer IV tank destroyers, and 1276b which received ten Sturmgeschütz III assault guns and was eventually attached to 180.Infanterie-Division.

4. This company appears in some sources as Sturmgeschütz- Kompanie 1326. Interestingly, a single Sturmgeschütz III is reported as being on hand in November but it is likely that this had been lost by the start of the Ardennes Offensive.

To my knowledge this photograph has not been published before and perhaps deserves to be included for that reason alone. Taken in April 1945 on the front of the US 3rd Armored Division, it shows two later production Hetzers. Although difficult to see here, the triangle-shaped braces welded to the front towing extensions, introduced in early 1945, are just visible on the vehicle in the foreground. At that time the American division was advancing through Paderborn to Nordhausen towards Dessau. Although it is not possible to identify the unit to which these vehicles belong with any certainty, possible candidates are Panzerjäger-Abteilungen 272, 277 and 1089, all of which had a number of Jagdpanzer 38 and are shown on 12th Army Group situation maps in the area during late March and early April 1945.

Panzerjäger-Kompanie 1352b on 27 November but the company did not operate with its parent formation and was instead attached to 79.Volksgrenadier-Division which was being rebuilt at that time. As part of 352.Volksgrenadier-Division the battalion took part in the Ardennes Offensive with just six serviceable tank destroyers and was reduced to little more that a weak Kampfgruppe by early 1945.

Panzerjäger-Abteilung 363. The battalion was formed in September 1944 at Truppenübungsplatz Wildflecken and attached to 363.Volksgrenadier-Division. This was one of a few Panzerjäger battalions to contain two assault gun companies and Sturmgeschütz-Abteilung 1363 was allocated fourteen Hetzers on 16 September.

Although the division's formation was incomplete it was rushed to Holland on 28 September 1944 to counter the Allied landings at Arnhem. But the Hetzers remained at Wildflecken. On 21 November 1944 another fourteen Hetzers were shipped to the battalion's Panzerjäger-Kompanie 1363b and by this time the division had been withdrawn from the fighting. At the end of November 363.Volksgrenadier-Division returned to the front and took part in the defence of the Jülich bridgehead, particularly around Hasenfeld, and the Ardennes Offensive as part of 15.Armee.

During early 1945 the division was engaged in the defence of western Germany, including the city of Cologne, but could only form a regimental-size Kampfgruppe and even the artillerymen were fighting as infantry. The division was trapped in the Ruhr Pocket and surrendered in April 1945.

Panzerjäger-Abteilung 416. Formed in November 1944 with three companies, the battalion was attached to 416.Infanterie-Division which had actually been fighting on the Western Front since the previous month. The battalion's Panzerjäger-Kompanie 1416 was allocated fourteen Hetzers on 1 January 1945 and these arrived in the same month.

The division fought mainly in the Saarpfalz region as part of 1.Armee and was defending the area between Mühlendorf and Wasserburg, west of Bamberg, on 1 May 1945 when it was almost completely destroyed by an American offensive.

...........text continued on page 12

Both photographs on this page were taken at Schwarzbach, modern-day Cerna v Posumavi in the Czech Republic, on 7 May 1945 when German units surrendered to the US Army. These Jagdpanzer 38 tank destroyers have been associated with 11.Panzer-Division in several sources but that formation was never allocated any Hetzers. As described in the unit histories, these vehicles may have been attached to an ad-hoc battalion-sized unit formed at Truppenübungsplatz Milowitz on 17 April 1945. Built around the staff of Panzerjäger-Abteilung 553, the battalion also contained Jagdpanzer-Kompanie 1159 and two Personaleinheiten units of company strength, PE 10d and PE 48c. Ordered to move towards the Western Front, the battalion's rail transport was stopped at Pilsen and the unit is last mentioned as travelling to Wallern, present-day Volary, which is less than 15 kilometres north of Schwarzbach. The white Hakenkreuz marking visible in the top photograph was added to a number of these vehicles by Allied soldiers and this Hetzer is actually the same tank destroyer on the left in the second image. Vehicles from what may be this unit are shown on page 24 of the Camouflage & Markings section.

............*text continued from page 10*

Notes

1. This unit had been part of Panzer-Brigade 112. It is unlikely that they had any of their own vehicles.
2. Curiously, an order dated 13 March 1945 directed that Panzerjäger-Abteilung 1089 was to be renamed Panzerjäger-Abteilung 89. This may have been related to the rebuilding of the division and it does not seem to have affected the Panzerjäger-Kompanie.
3. A Wehrkreis was a military administrative district.

Panzerjäger-Abteilung 553. See Panzerjäger-Abteilung 1553 below.

Panzerjäger-Abteilung 708. Formed in September 1944, the battalion was attached to 708.Volksgrenadier-Division. On 23 October a total of fourteen Hetzers were shipped to the battalion and these were reported as being on hand with Panzerjäger-Kompanie 1708, the assault gun company, by November.

The battalion was virtually destroyed in the fighting in Alsace in January 1945 and the survivors of were used to form the headquarters element of Panzerjäger-Abteilung zbV 'G' (see Panzerjäger-Kompanie Bock below).

The division was withdrawn from the front on 13 February 1945 and plans to rebuild it were later abandoned. A vehicle of the unit is shown on page 17 of the Camouflage & Markings section of this book.

Panzerjäger-Abteilung 711. Attached to 711.Infanterie-Division, the battalion was almost destroyed in the Normandy battles and was withdrawn to Holland and rebuilt with three companies. In November 1944 a total of fourteen Hetzers were shipped to the battalion and these had arrived by the following month. Very soon after this the complete division was transferred to Hungary and remained on the Eastern Front until the end of the war.

Panzerjäger-Abteilung 716. Formed in September 1944, the battalion was attached to 716.Infanterie-Division. The battalion's assault gun company, Panzerjäger-Kompanie 1716, was allocated fourteen Hetzers on 25 November 1944 and these were reported as being on hand in December.

The division was destroyed in the fighting for the Colmar Pocket in January 1945 shortly after the Panzerjäger battalion was allocated four new Hetzers. The division was rebuilt in April 1945 but Panzerjäger-Kompanie 1716, which had been training at Milowitz, never rejoined its parent formation and was attached to Panzerjäger-Abteilung Lambert and ended the war fighting in central Germany. See Panzerjäger-Abteilung Lambert below.

Panzerjäger-Abteilung 1089. The battalion was raised in early October 1944 for 189.Infanterie-Division and was made up of three companies. The battalion's second company was equipped with Sturmgeschütz III assault guns and was referred to as a Sturmgeschütz-Batterie. In January 1945 the company's personnel were transferred to Milowitz in Czechoslovakia and replaced by Sturmgeschütz-Kompanie 2112 (1) and in the following month the division was destroyed in the battles in Alsace. The men of the battalion's second company, now renamed Panzerjäger-Kompanie 1089, were allocated ten Hetzers and one Bergepanzer 38 recovery vehicle in mid-March 1945 and as the attempt to reform the division was abandoned in April they were directed to travel from Milowitz to the Western Front where the company ended the war as part of Panzerjäger-Abteilung Lambert fighting in the Harz Mountains (2). See Panzerjäger-Abteilung Lambert below.

Panzerjäger-Abteilung 1316. Attached to 16.Volksgrenadier-Division, the battalion was formed in October 1944 from the remnants of Sturmgeschütz-Abteilung 1316 and Panzerjäger training units of Wehrkreis V (3).

The division was sent to the front after a short period of training, operating in the Vosges region in the defensive fighting between Langres and Epinal as part of 19.Armee, but the Panzerjäger battalion did not receive its first allocation of Hetzer tank destroyers until early December 1944. These were all assigned to the battalion's second company which was referred to as Panzerjäger-Kompanie 1316.

In early 1945 the division was defending the front line between Katzenthal and Wintzenheim, west of Colmar, but was in danger of being encircled and in the first week of February it was withdrawn to the east bank of the Rhine near Neuenburg. In March 1945 the division was fighting around Karlsruhe but had been split into a number of Kampfgruppen. On 21 April 1945 the division was officially disbanded.

Panzerjäger-Abteilung 1553. The battalion's parent formation, 553.Volksgrenadier-Division, was formed in October 1944 and destroyed and rebuilt, only to be destroyed once more before the end of the year.

The Panzerjäger battalion was rebuilt in February 1945 and the assault gun company, Panzerjäger-Abteilung 1553, was allocated ten Hetzers and one Bergepanzer 38 by the HZA which were shipped on 26 March 1945. However, the division had returned to the front two days earlier, fighting around Würzburg and Crailsheim, and it is likely that the Panzerjäger battalion remained at Milowitz where it was training. At about this time the battalion was renamed Panzerjäger-Abteilung 553.

On 17 April 1945 a number of Gemischte, or mixed, Panzerjäger formations were formed from the units at Milowitz and one of these contained the staff of Panzerjäger-Abteilung 553, Jagdpanzer-Kompanie 1159 and two Personaleinheiten units of

Above: Two pages from a US Army report on German armoured vehicles show the early production Hetzer of Heeres-Panzerjäger-Abteilung 741 depicted in the Camouflage & Markings section on page 17 of this book. This battalion was the first Hetzer unit to go into action on the Western Front and this vehicle may have been the earliest example captured by the Allies. Although faded, the photographs of the Rundumsfeuer remotely-controlled machine-gun mount show clearly how the gun was attached. Note also the Pilze socket immediately behind the gun bracket, next to the loader's hatch.

company strength, PE 10d and PE 48c (1). These last three units had originally been formed for the first three Reichsarbeitsdienst (RAD) divisions (see below). The vehicles allocated to these units may have been the last Hetzers shipped from the HZA before the end of the war.

The unit left Milowitz during the evening of 24 April 1945 by rail for the Western Front but just two days later the train was halted near Pilsen and the Hetzers were ordered to move to Wallern, north-east of Passau, on their own tracks. There is no evidence that this unit ever engaged in combat and it is possible that these are the vehicles that surrendered with elements of 11.Panzer-Division at Schwarzbach on 7 May 1945.Vehicles from what may be this unit are shown on page 24 of the Camouflage & Markings section.

Panzerjäger-Abteilung 1560. The battalion was formed in November 1944 in Norway with three companies and attached to 560.Volksgrenadier-Division. On 19 November 1944 a total of fourteen Hetzers were shipped to the battalion by the HZA and these were assigned to Panzerjäger-Kompanie 1560. The division took part in the Ardennes Offensive as part of LVIII.Panzerkorps but it seems that none of the Hetzers were available when the attack began, although they were involved in the later fighting. The American counterattack in the St Vith sector hit the division particularly hard as elements were still arriving from Norway. In February 1945 the division was involved in the fighting around Echternach, on the Luxembourg frontier north-west of Trier, and was subsequently pushed back across the Pruem River towards Bitburg. The division was trapped in the Ruhr Pocket and completely destroyed in April 1945.

Panzerjäger-Abteilung 1818. The battalion was formed in September 1944 from elements of the partially-raised Panzerjäger-Abteilung 1571. Attached to 18.Volksgrenadier-Division, the battalion was allocated fourteen Hetzers on 22 October 1944. These were on hand by November and all were allocated to the battalion's second company which was referred to as Sturmgeschütz-Abteilung 1818. The division took part in the Ardennes Offensive in December 1944 capturing the town of St Vith but suffered heavy losses in the process. For the remainder of the year the division was involved in defensive operations but by the middle of January it was thrown into the fighting around Bleialf, just inside German territory, although the grenadier

...........text continued from page 15

Notes

1. Personaleinheiten were replacement units made up of personnel only with no equipment.

Photographed near Halloville, about 20 kilometres east of Luneville in north-eastern France, this Hetzer is one of the vehicles of Panzerjäger-Abteilung 708 which was involved in the fighting there. On the morning of 14 November 1944, elements of the US 79th Division, advancing from Montigny south of Nancy, reached the village of Halloville where the defensive lines of 553.Volksgrenadier-Division and 708.Volksgrenadier-Division, commanded by Generalmajor Josef Krieger, met. Immediately realising the danger, Krieger ordered his men to prepare to counterattack but as the German units were assembling to the east of the village church they were struck by a combined American force supported by tanks and tank destroyers. Caught completely by surprise, the Grenadiers and their supporting assault guns of Panzerjäger-Abteilung 1708 were forced into a disorderly retreat, losing most of their armoured vehicles in the process. This vehicle and its intricate factory-applied camouflage scheme is also shown and discussed on page 17 of the Camouflage & Markings section.

HEERES-PANZERJÄGER-ABTEILUNG, AUGUST 1944

A number of independent Panzerjäger battalions had served with the army since the beginning of the war, equipped with a variety of towed and self-propelled anti-tank weapons. These units had proved successful and in July 1943 orders were issued by the Oberkommando des Heeres (OKH) for the creation of a series of fully-armoured Panzerjäger battalions which were to be subordinated to Heeresgruppe commands on the Eastern Front. These units were all to be numbered in the 700 series and by November 1943 Panzerjäger-Abteilungen 731, 741, 743 and 744, which were all eventually equipped with the Hetzer, and Panzerjäger-Abteilungen 721, 742, 751 and 753 were operational. Of these units only Panzerjäger-Abteilung 741 served in the West. The diagram shown here is based on orders issued for the formation of Panzerjäger-Abteilung 731 in May 1944 which is the best surviving evidence we have for the organisation of these units and it is reasonably safe to assume that the Western Front battalion would have been essentially the same.

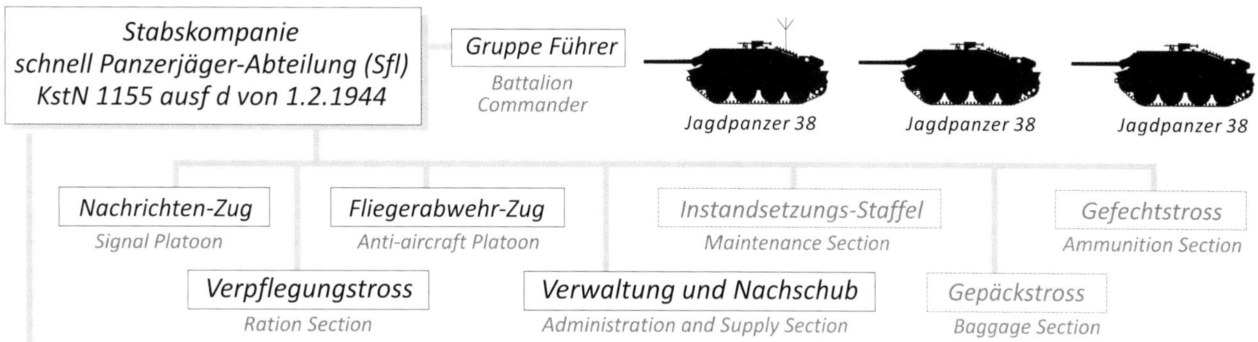

The OKH order of 25 May 1944 included the notation: Stab. (ohne Gepäcktross, zusätzlich J-Staffel, Gef. Tross usw. Aus Stabs-Kp. 1155 d v.1.11.43 nach anliegender gekurzter Gliederung)(nur and Chef H Rüst u Bbk/AHA/Stab I). Meaning, that the Staff element was to be organised using an abbreviated version of Stabs-Kompanie, schnell Panzer-Jäger-Abteilung (Sfl) KstN 1155 ausf d von 1.11.1943 without the Baggage, Maintenance and Ammunition Sections which are here shown in dark grey. These services were to be provided by the higher formation to which the battalion was subordinated. It has also been suggested that these battalions lacked the Fliegerabwehr-Zug with its three Sdkfz 7/1 halftracks mounting 2cm Flakvierling 38 anti-aircraft guns. Similarly the companies were built with an amended KstN which did not allow for any repair or recovery facilities. This apparently proved unworkable as the Eastern Front units were all issued Bergepanzer 38 recovery vehicles. There is evidence that Heeres-Panzerjäger-Abteilung 741 was organised with a complete headquarters element using KstN 1155 ausf d or even KstN 1106. In any case its fighting companies would have been organised as shown below.

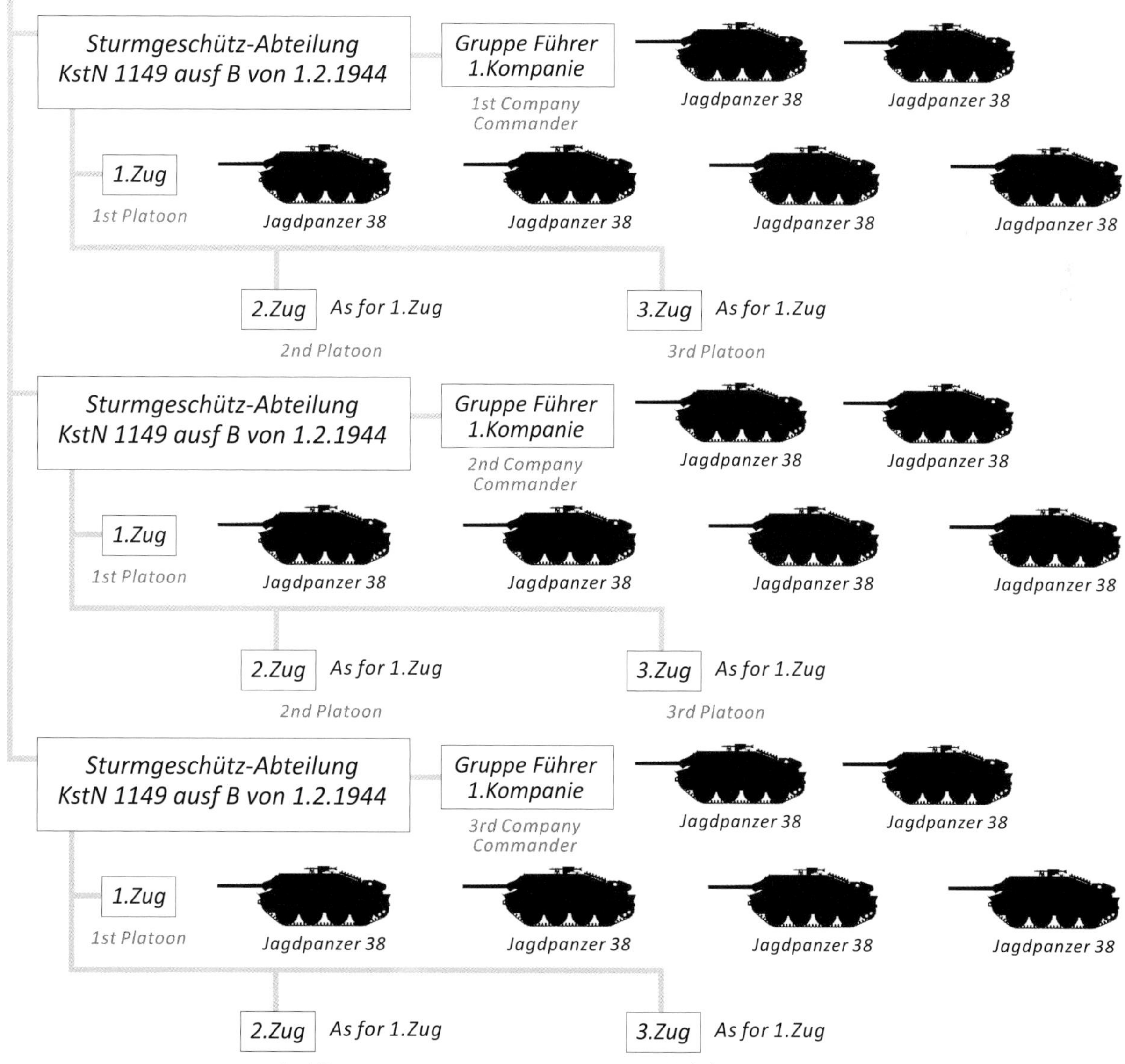

...........text continued from page 13

regiments were almost completely exhausted. In March the survivors fell back across the Rhine and were combined with the remnants of 26.Volksgrenadier-Division to form an ad-hoc Kampfgruppe. A vehicle of this unit is shown on page 19 of the Camouflage & Markings section.

Panzerjäger-Abteilung Potsdam. Attached to Infanterie-Division Potsdam, the battalion was raised on 4 April 1945 from elements of Panzerjäger-Abteilung 85 which had been badly depleted in the fighting around Aachen in the previous month.

Like most late war formations this division was vastly understrength and the Panzerjäger battalion may have consisted of just a single company, Jagdpanzer-Kompanie Potsdam, formed from elements of Panzerjäger-Kompanie 1185. Ten Hetzers and one Bergepanzer 38 were shipped by the HZA on 12 April, but it is not clear if these ever arrived at their destination.

The division was almost completely destroyed in the fighting in the Harz mountains and the survivors were used to form the Füsilier-Bataillon of Infanterie-Division Scharnhorst.

Panzerjäger-Abteilung Scharnhorst. Formed in late March and early April 1945, the battalion was attached to Infanterie-Division Scharnhorst.

The battalion was allocated ten Hetzers and one Bergepanzer 38 on Sunday, 8 April 1945 and the division's commander was ordered to have his formation ready for combat by the following Thursday. The battalion's assault gun company was referred to as Jagdpanzer-Kompanie Scharnhorst.

The division was initially deployed at Barby on the Elbe south-east of Magdeburg against units of the US Army and then moved to the Eastern Front, which was by that time just 60 kilometres away, during the last days of the month. The survivors of the division surrendered on 2 May 1945 at Tangermünde on the Elbe River.

Panzerjäger-Abteilung Ulrich von Hutten. Attached to Infanterie-Division Ulrich von Hutten, the battalion was made up of a staff element and three full companies including Jagdpanzer-Kompanie Ulrich von Hutten (Gemischte) which was allocated ten Hetzers and one Bergepanzer 38 on 10 April 1945 (1). the company also contained a Grenadier-Begleit platoon.

As part of 12.Armee the division took part in the attempt to relieve the garrison of Berlin and then retreated towards the West, surrendering to units of the US Army at Tangermünde on 8 May 1945.

Panzerjäger-Abteilung Lambert. This independent formation was raised by an order of 30 March 1945 and contained the headquarters staff of Panzer-Abteilung 44 and Jagdpanzer-Kompanien 1106, 1089 and 1716. Each of the three companies had been equipped with ten Hetzers and one Bergepanzer 38 in the previous weeks at Milowitz.

...........text continued on page 49

Notes

1. This was actually the second incarnation of the division as the original formation was destroyed in the defence of Heiligenbeil in East Prussia. Only the battalion's second company was referred to by the division's name. As a matter of interest, the other companies were 1.(Panzer-Zerstörer)Kompanie (Gemischte) and 3.(Flak) Kompanie (Gemischte).

Photographed in the streets of Kesternich, a village north-east of Monschau in western Germany, in late January 1945, the camouflage applied to these two Hetzers is indicative of the confused state of affairs at this stage of the war. While the vehicle in the background has received a very comprehensive coat of whitewash camouflage, the Jagdpanzer on the right seems to have been painted in a single dark colour. Both are fitted with the early cast driver's visor suggesting that they were assembled after September 1944. If the location and date are correct, these vehicles may be from either 62.Volksgrenadier-Division or, more likely, the Panzerjäger battalion of 272.Volksgrenadier-Division which is shown just outside Kesternich on the OKW situation map for 1 January 1945.

1. Jagdpanzer 38. Unit uncertain. Holland, September 1944. The first Hetzers deployed to the Western Front were allocated to the Panzerjäger battalion of 183.Volksgrenadier-Division and the independent Heeres-Panzerjäger-Abteilung 741 and this vehicle is probably from the latter.

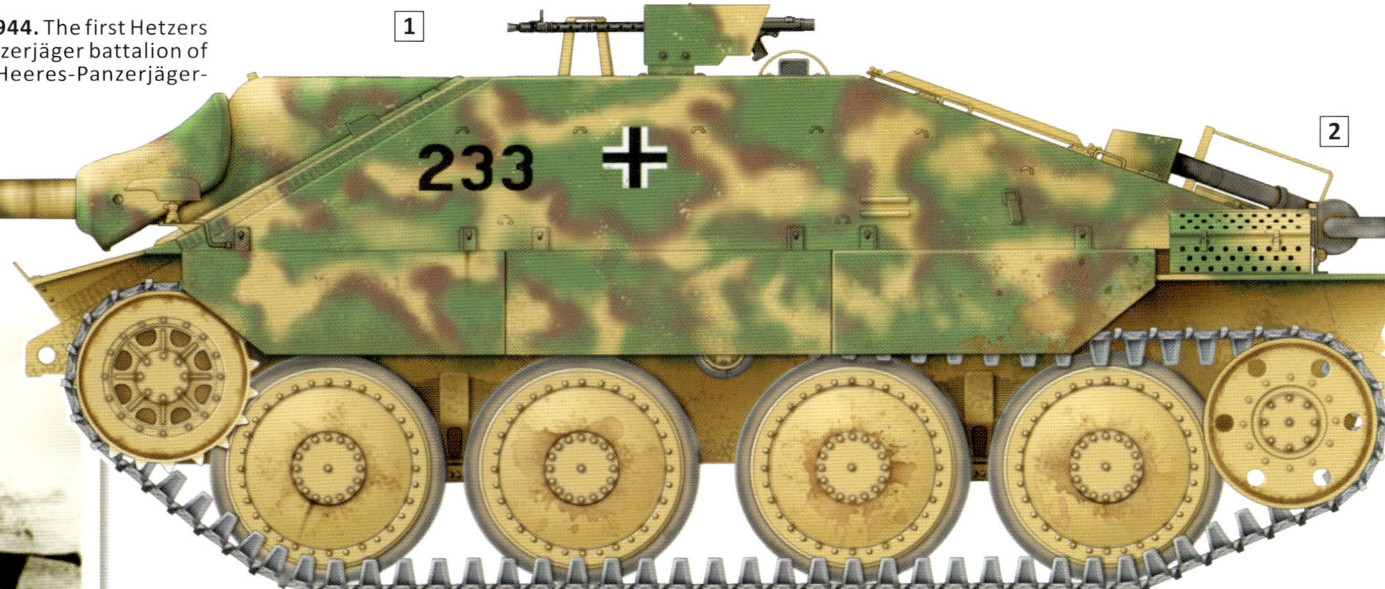

Above: This Skoda-built vehicle has the six-hole flat rear idler and the early 32-bolt roadwheels. Note the field-modified rack on the rear hull (2), also visible in our photograph (3). The unit insignia (4), just visible in our photograph, is highly conjectural on my part and is in fact based on the formation badge of Sturmgeschütz-Abteilung 277.

5. Jagdpanzer 38. 708.Volksgrenadier-Division. France, December 1944. The division reported that fourteen of these tank destroyers were on hand from November and this vehicle was disabled in a brief firefight near Halloville, east of Nancy. It is painted in what may be one of the first factory-applied camouflage schemes employed by BMM. Note that the small wedges of contrasting colour do not seem to follow any pattern and could have been applied individually. Interestingly, the inside of the machine-gun shield is also painted in the same style. The company number is depicted here as a faded black but could also have been rendered in another dark shade. Gun barrels were coated in a black heat-resistant laquer and replacements were commonly left unpainted, as here.

1

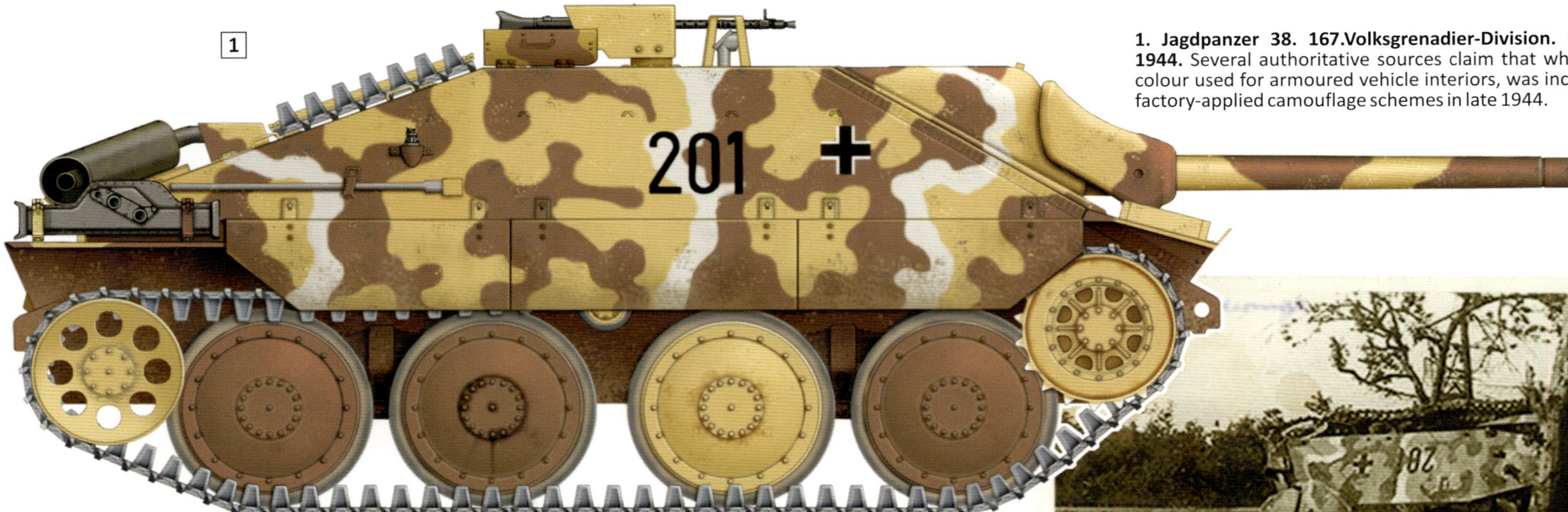

1. Jagdpanzer 38. 167.Volksgrenadier-Division. Belgium, December 1944. Several authoritative sources claim that white or Elfenbein, the colour used for armoured vehicle interiors, was incorporated into some factory-applied camouflage schemes in late 1944.

2

Above: While the use of white as a camouflage colour is the subject of debate, with some researchers claiming the lighter shade visible in our photograph (2) is Dunkelgelb, I feel there is a case for both arguments and so I have rendered the shade here as a very faded white as it is clear that manufacturers were thinning the paint at this time. The possible use of white paint as a camouflage colour is also discussed on page 21.

3. Jagdpanzer 38. 257.Volksgrenadier-Division. Belgium, December 1944. Finished in the factory-applied camouflage scheme associated with BMM-assembled vehicles this Hetzer is from Panzerjäger-Kompanie 1179 which replaced the division's assault gun company in late November or early December 1944. The loading label (4) was in theory painted onto all vehicles, including trucks and cars, and indicated the type, weight and loading class. The latter, in the case of fully-tracked vehicles, was class II rendered as Ve.Kl.II. The notation could vary from basic to quite detailed (5).

3

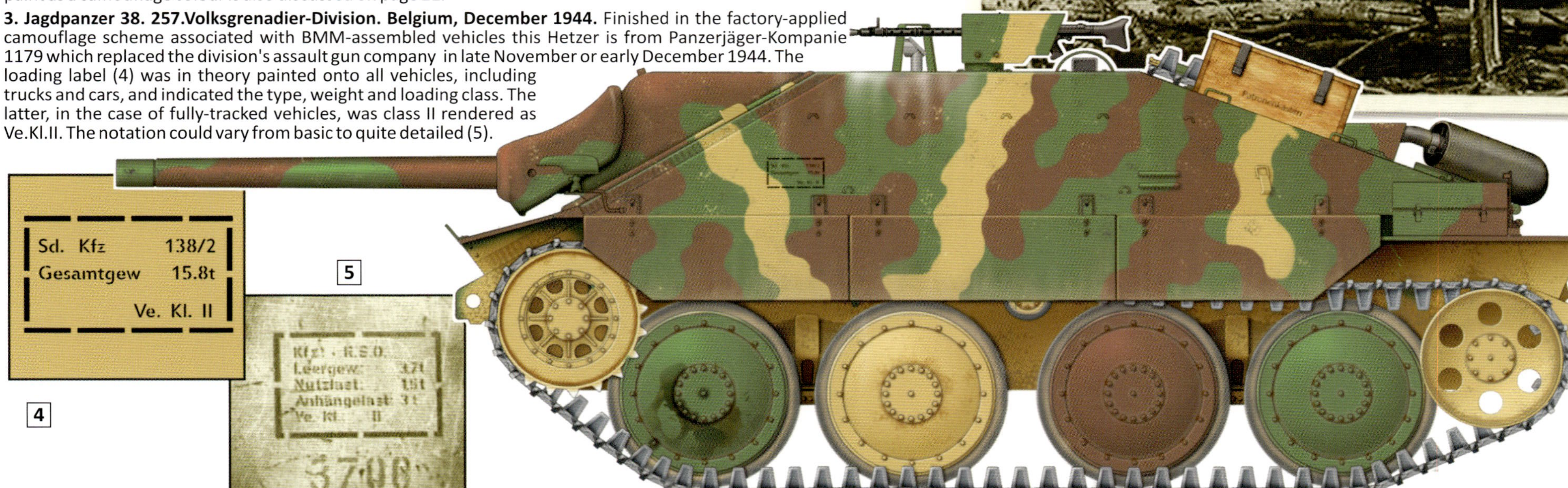

4

5

1. Jagdpanzer 38. Unit uncertain. Germany, January 1945. Disabled near Monschau in western Germany, this Hetzer is finished in a neat coat of whitewash camouflage, which seems to have been quite unusual. This vehicle was fitted with the early cast driver's visor but it was shot away, as can be seen from our photograph (2).

1

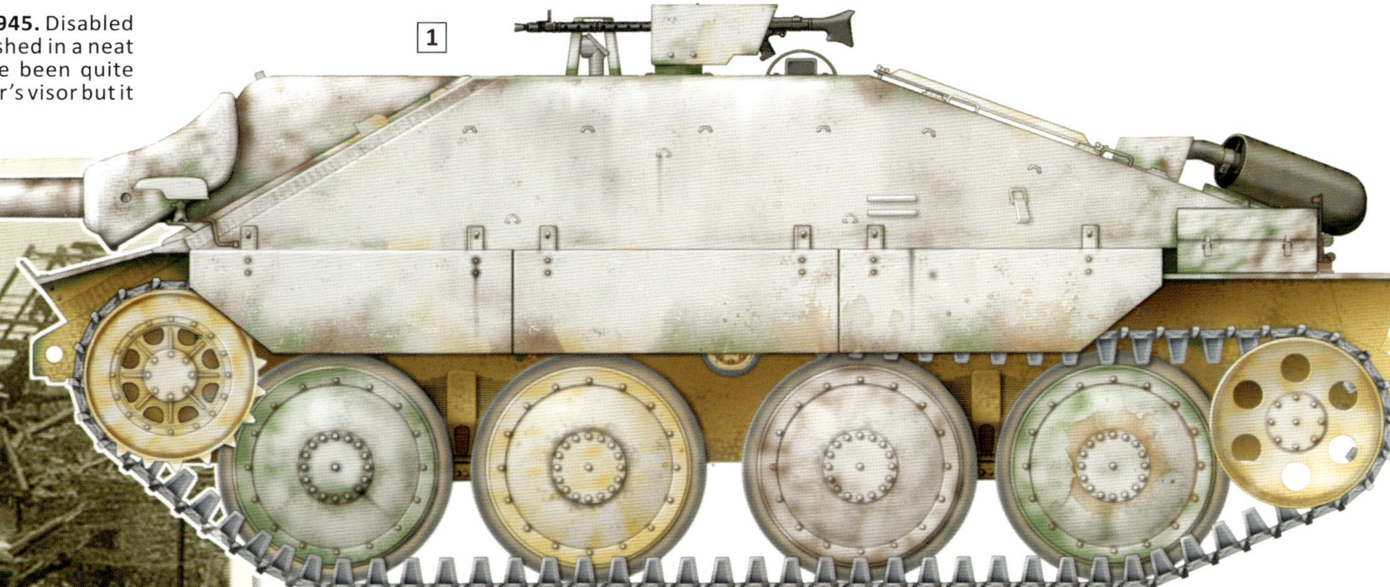

2

Above: The six-hole rear idler and Flammvernichter muffler are speculative on my part but are correct for the period. This Hetzer is probably from the Panzerjäger battalion of 272.Volksgrenadier-Division and a larger image is reproduced on page 16.

3

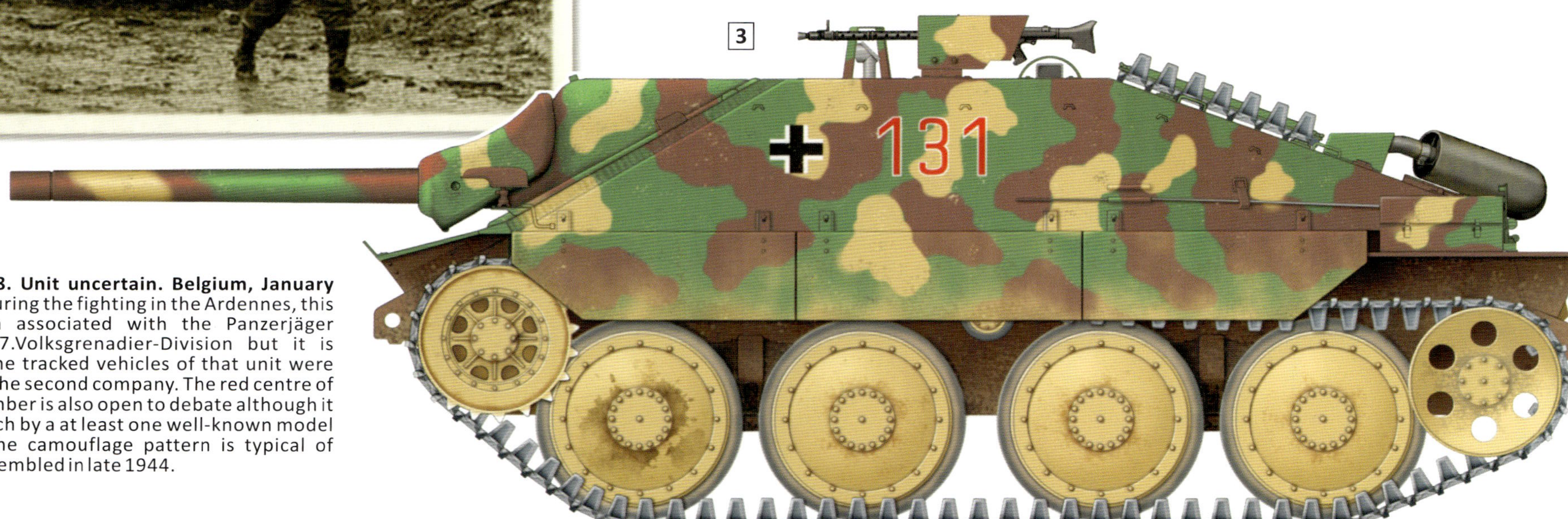

3. Jagdpanzer 38. Unit uncertain. Belgium, January 1945. Pictured during the fighting in the Ardennes, this vehicle is often associated with the Panzerjäger battalion of 167.Volksgrenadier-Division but it is known that all the tracked vehicles of that unit were concentrated in the second company. The red centre of the company number is also open to debate although it is rendered as such by a at least one well-known model manufacturer. The camouflage pattern is typical of BMM Hetzers assembled in late 1944.

1. Flammpanzer 38. Flammpanzer-Kompanie 353. France, December 1944. Just twenty Hetzers were converted to the flamethrowing role and allocated to the two companies mentioned here. The thin metal cover for the flamethrower was easily damaged and the front section is missing entirely from this Flammpanzer. Our photograph (2) shows the flame projector and the cover which was intended to disguise the tank's true purpose. All the vehicles of this company were numbered as shown here, with the number preceded by a letter S. The yellow colour is speculative but entirely possible. Photographs of this vehicle are reproduced on page 51 and page 53.

1

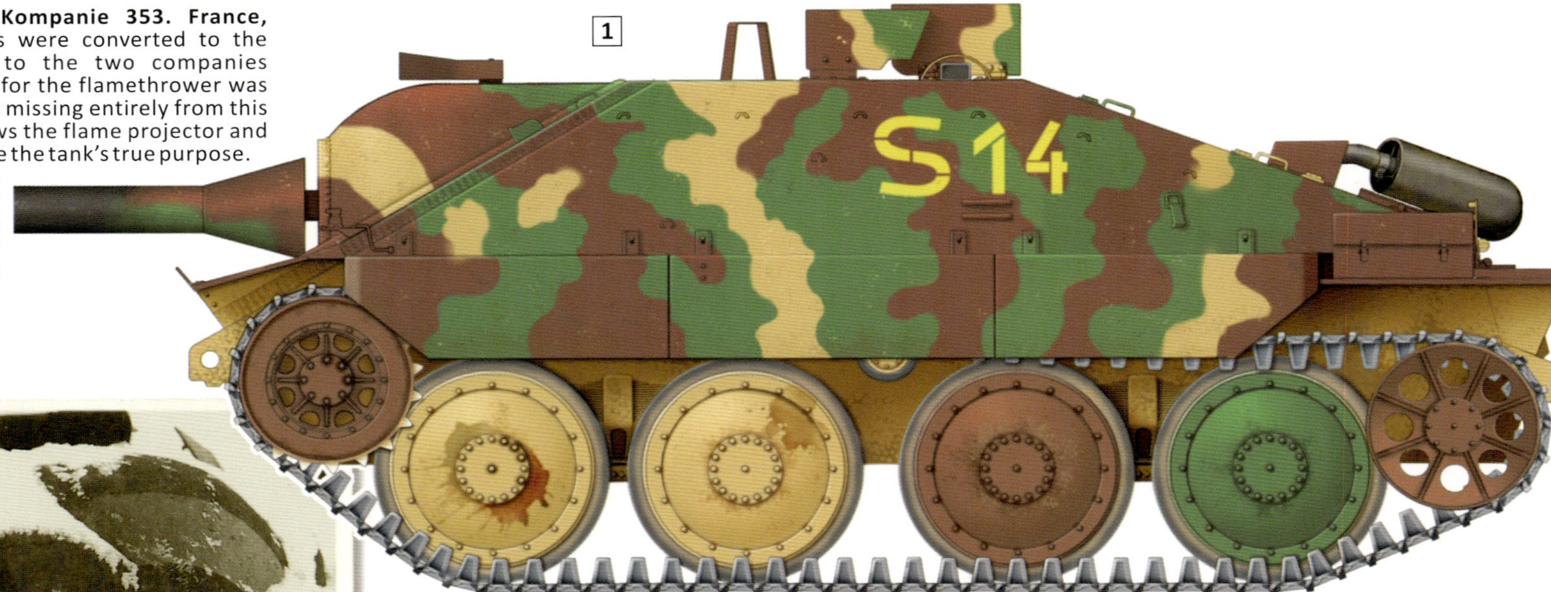

2

3. Flammpanzer 38. Flammpanzer-Kompanie 352. Germany, spring 1945. This vehicle is quite unusual in that it is fitted with the metal foliage rings on the hull side, which were introduced into production in early 1945, and also seems to have been finished in factory-applied paint scheme typical of Skoda-assembled tanks. This is discussed further on page 52 where a photograph of this vehicle can be seen.

3

1

3

1. and 2. Jagdpanzer 38. 600(Russ).Infanterie-Division. Czechoslovakia, spring 1945. Although this unit served in the East until the last days of the war when the survivors attempted to surrender to the US Army, I have included these vehicles here to illustrate the camouflage scheme employed by BMM in 1945.

2

As mentioned in the Technical Details and Modifications section, a number of authoritative sources have suggested that thinned white paint, or possibly Elfenbein RAL 1001, was used as a camouflage colour for a period in December 1944. Although this light colour could be Dunkelgelb RAL 7028, some black and white photographs do seem to show two separate light shades and I have presented two possibilities here, both based on photographs. It should be mentioned that the use of four-colour camouflage can clearly be seen in photographs of Sturmgeschütz IV assault guns taken at about this time. Compare the pattern with our photograph (3) which shows a Hetzer from the same unit in Prague in April 1945.

1. Jagdpanzer 38. 600(Russ).Infanterie-Division. Czechoslovakia, spring 1945. From the same battalion as the vehicles depicted on the previous page, this final production Hetzer is included here to show what may have been the last factory-applied camouflage scheme employed by BMM. Also shown is the division's formation insignia (2). The Cyrillic letters stand for Russkaya osvoboditel'naya armiya, or Russian Liberation Army.

1

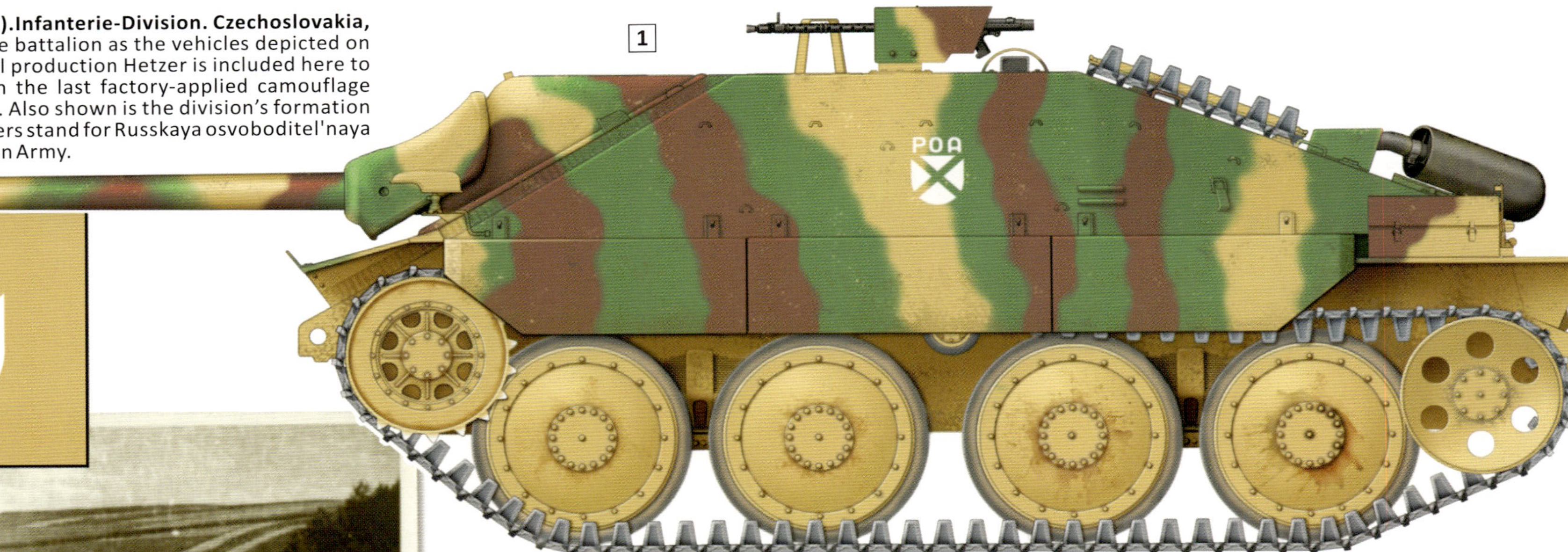

2

4

3

3. Jagdpanzer 38. Unit uncertain. Germany, spring 1945. Many early production Hetzers, such as the vehicle shown here, were still in service when the war ended in May 1945. The first pattern gun mantlet, rivetted hull Schürzen and field-applied camouflage would suggest that this tank destroyer was assembled at some time before mid-August 1944. Unusually, the camouflage seems to have been extended to the roadwheels as can be seen from our photograph (4). Note the white band on the barrel of the main gun and the large Balkenkreuz national insignia.

1

2

3

1. Jagdpanzer 38. Unit unknown. Czechoslovakia, spring 1945. Note that the front extension towing eyes of this early production vehicle have been broken or cut away and replaced with a simple iron rod (2). The naming of armoured vehicles (3) was not widespread in the German army but was apparently common within this battalion with at least one other example, Babett, being known. Note that this Hetzer is fitted with an MG 42.

4. Jagdpanzer 38. 97.Jäger-Division. Czechoslovakia, spring 1945. The Flammvernichter exhaust muffler, solid toolbox and 16-bolt roadwheels would date this vehicle's assembly to some time after September 1944. The camouflage pattern was almost certainly applied at the assembly plant and is quite similar in style to that seen on the Hetzer at the top of page 25. It is quite unlike the patterns seen in photographs of both BMM and Skoda vehicles built in late 1944 and may have been very short-lived. Note also the two Balkenkreuz insignia.

4

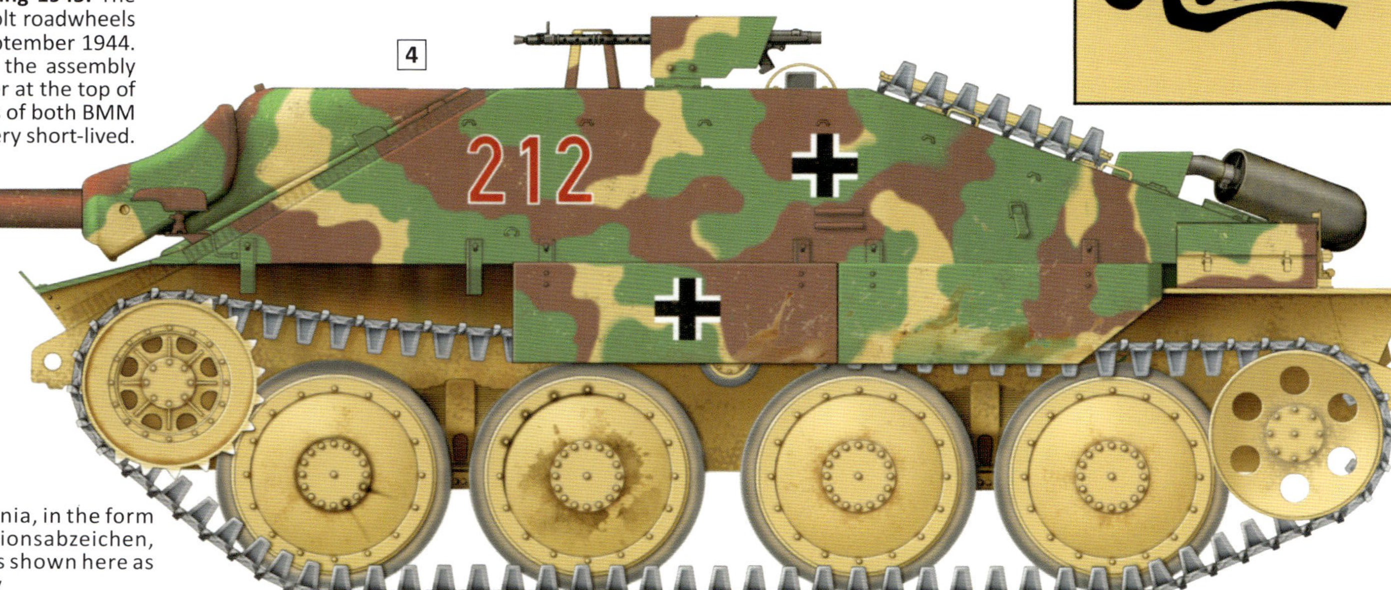

5

The division's unit insignia, in the form of an unofficial Traditionsabzeichen, or tradition badge (5), is shown here as a matter of interest only

1. and 2. Jagdpanzer 38. Unit uncertain but possibly Gemischte Panzerjäger-Abteilung Stab 553. Germany, spring 1945.

Note the large wooden box attached to the rear hull of both vehicles. This seems to have been a common feature of the Hetzers photographed at Schwarzbach. Although the images on which our illustrations are based are not perfectly clear, many vehicles being obscured by foliage camouflage, the number 23 shown on the hull side of the Hetzer above at left is just discernible. It is also obvious that the suspension of the Befehlspanzer shown below has not been painted in camouflage colours. Photographs of these tank destroyers are shown on page 11.

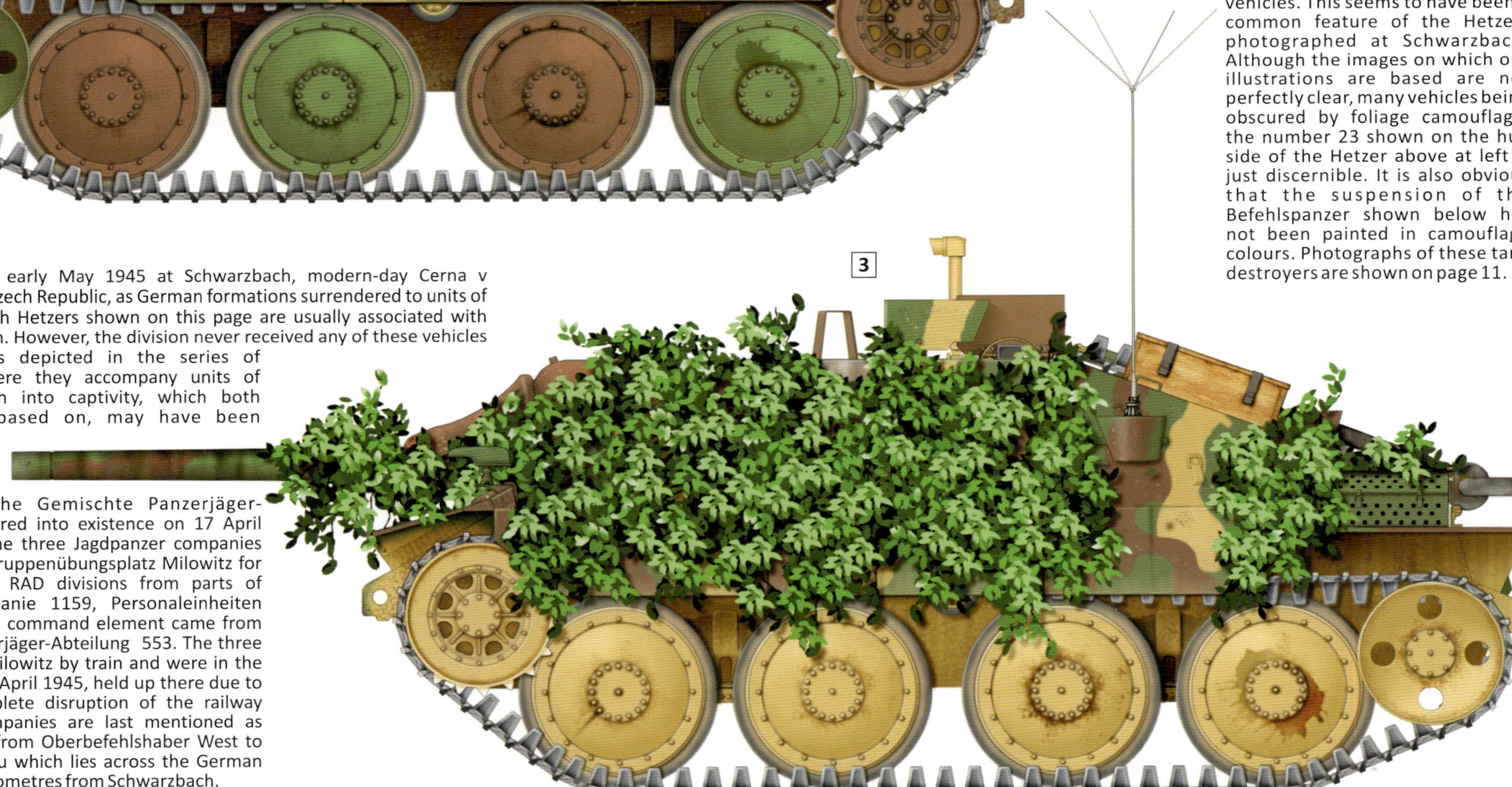

Photographed in early May 1945 at Schwarzbach, modern-day Cerna v Posumavi in the Czech Republic, as German formations surrendered to units of the US Army, both Hetzers shown on this page are usually associated with 11.Panzer-Division. However, the division never received any of these vehicles and the Hetzers depicted in the series of photographs where they accompany units of 11.Panzer-Division into captivity, which both profiles were based on, may have been from one of the Gemischte Panzerjäger-Abteilungen ordered into existence on 17 April 1945. At that time three Jagdpanzer companies were formed at Truppenübungsplatz Milowitz for the newly-raised RAD divisions from parts of Jagdpanzer-Kompanie 1159, Personaleinheiten 10d and 48c. The command element came from the staff of Panzerjäger-Abteilung 553. The three companies left Milowitz by train and were in the Pilsen area on 26 April 1945, held up there due to the almost complete disruption of the railway system. The companies are last mentioned as receiving orders from Oberbefehlshaber West to proceed to Passau which lies across the German frontier just 20 kilometres from Schwarzbach.

1

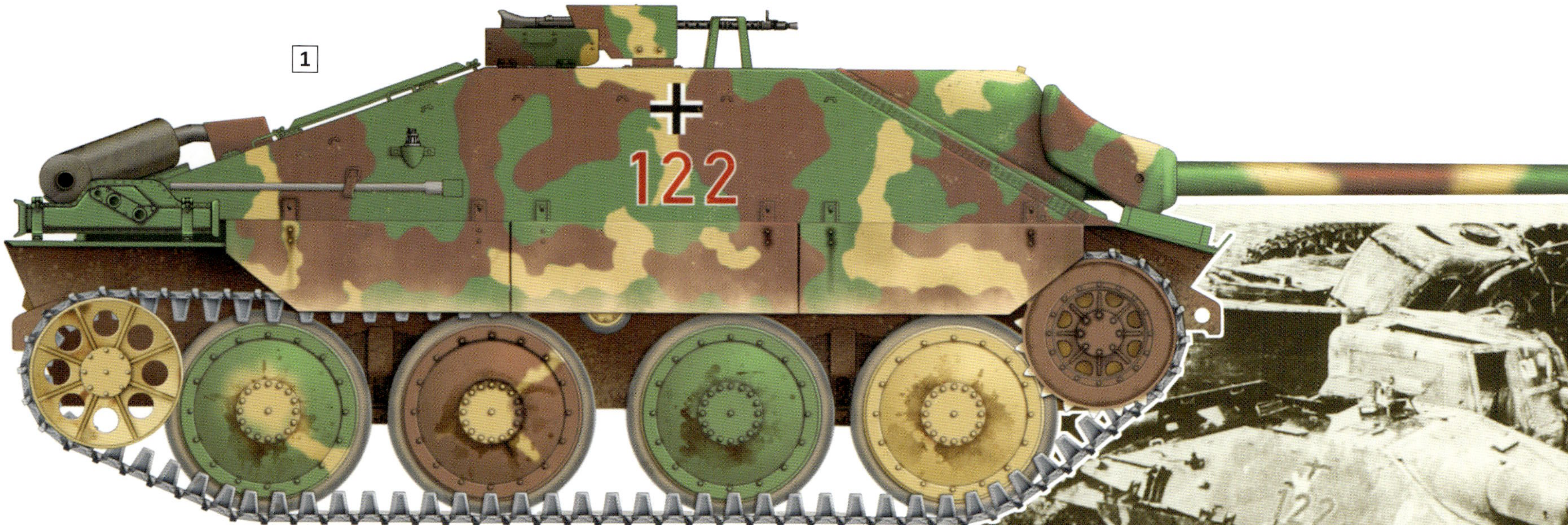

2

1. Jagdpanzer 38. 65.Infanterie-Division. Italy, April 1945. This rather complex camouflage pattern is extended to the roadwheels as can be seen from our photograph (2). Note the ribbed rear idler and the metal cover which closes off the Flammvernichter exhaust muffler. This latter may have been a modification which was unique to this battalion. Panzerjäger-Abteilung 165 was unusual in having its tracked vehicles concentrated in 1.Kompanie.

3

3. Jagdpanzer 38. 305.Infanterie-Division. Italy, April 1945. This Hetzer is painted in the camouflage pattern increasingly employed by BMM from September 1944 where the bands of Dunklegelb RAL 7028 gradually became thinner as seen here. A photograph of this vehicle is reproduced on page 62 of the Technical Details and Modifications section of this book.

1

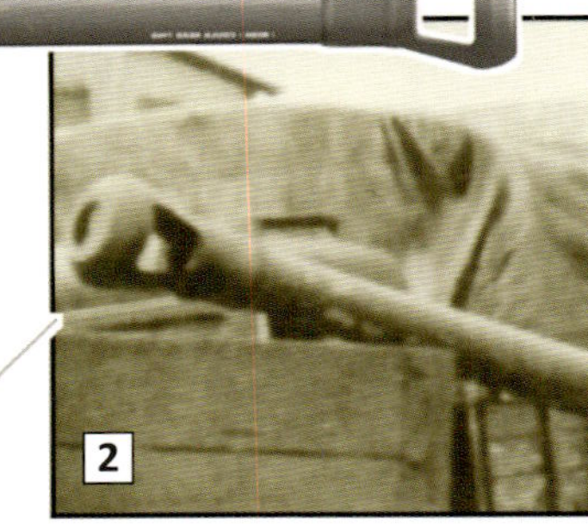

2

3

5

1. Jagdpanzer 38. Germany, May 1945. Photographed at Heeresversuchsanstalt Hillersleben, the army's massive testing facility north of Magdeburg, this Hetzer is fitted with a number of early production features including the gun mantlet, flat rear idler, 32-bolt roadwheels, cast exhaust cover and Schürzen attachments. Rheinmetall experimented with a number of guns and mounts based on the Jagdpanzer 38 at Hillersleben and this vehicle appears to be armed with a British OQF 75mm gun. The distinctive muzzle brake is clearly visible in our photographs (2 and 3). It is not, as has been suggested elsewhere, a 7.5cm KwK 40 L/43 gun as fitted to the so-called Pzkpfw IV ausf F2.

6

4

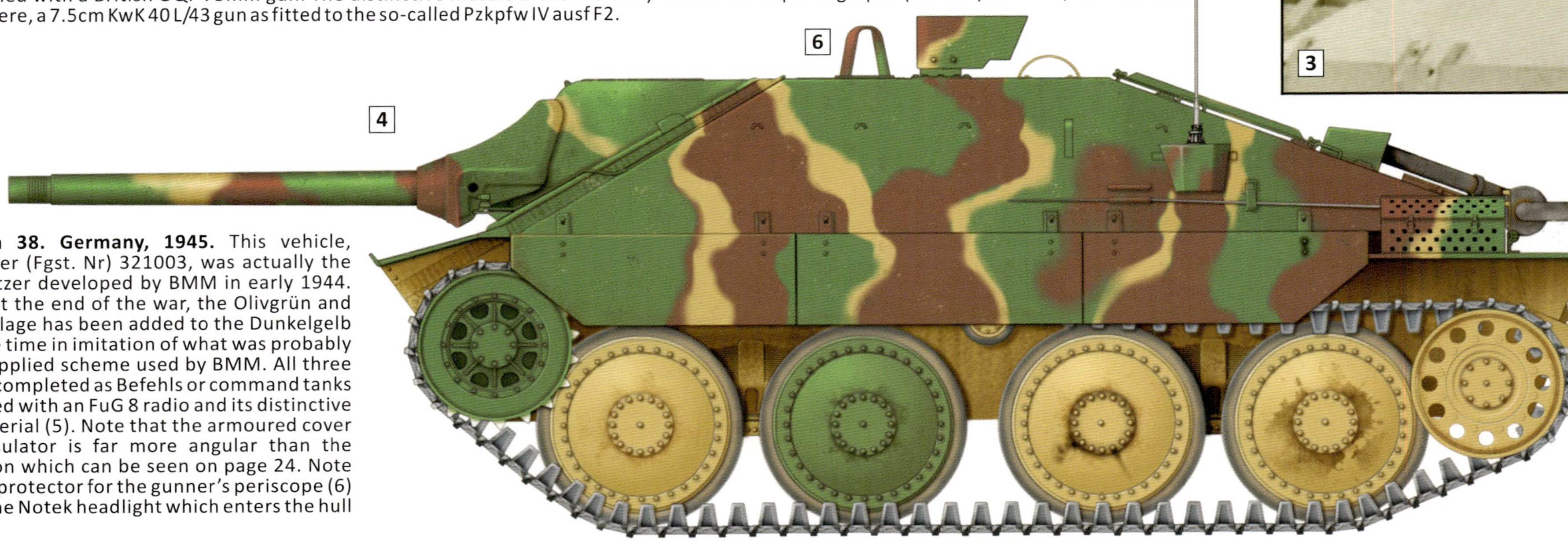

4. Befehlswagen 38. Germany, 1945. This vehicle, Fahrgestellnummer (Fgst. Nr) 321003, was actually the third protype Hetzer developed by BMM in early 1944. Captured intact at the end of the war, the Olivgrün and Rotbraun camouflage has been added to the Dunkelgelb base coat at some time in imitation of what was probably the last factory-applied scheme used by BMM. All three prototypes were completed as Befehls or command tanks and were equipped with an FuG 8 radio and its distinctive Sternantenna D aerial (5). Note that the armoured cover of the aerial insulator is far more angular than the production version which can be seen on page 24. Note also the rounded protector for the gunner's periscope (6) and the lead for the Notek headlight which enters the hull through the roof.

JAGDPANZER 38 HETZER

1/48 SCALE

TORLAP INTARARANGSON

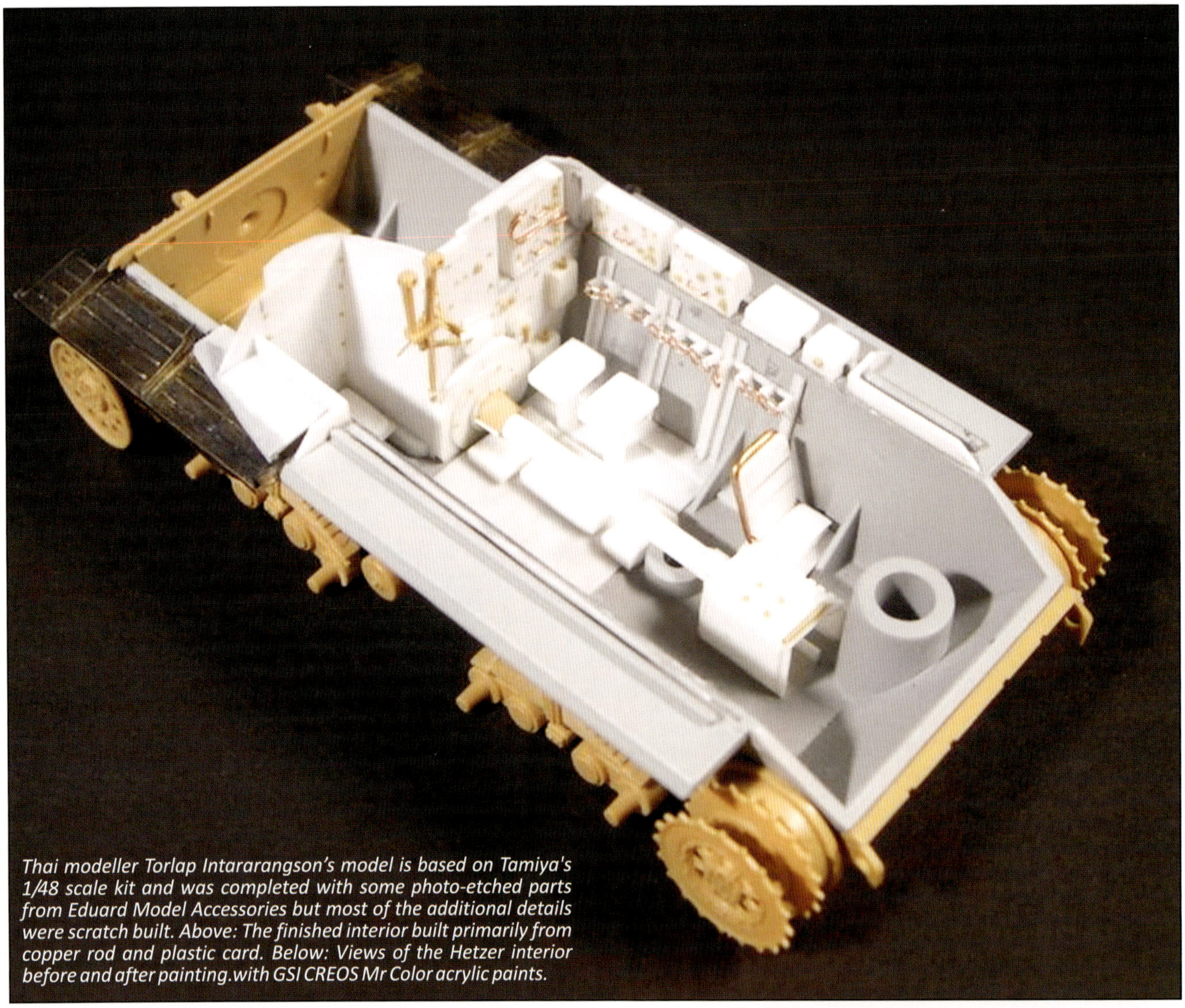

Thai modeller Torlap Intararangson's model is based on Tamiya's 1/48 scale kit and was completed with some photo-etched parts from Eduard Model Accessories but most of the additional details were scratch built. Above: The finished interior built primarily from copper rod and plastic card. Below: Views of the Hetzer interior before and after painting.with GSI CREOS Mr Color acrylic paints.

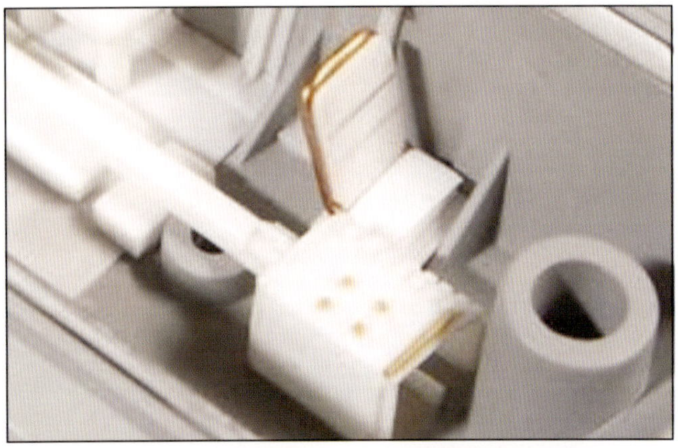

The finished build, less the tracks, ready for painting. The commander's and loader's hatches were left open to show the detailed interior. The exterior details were created using photo-etched details from Eduard Model Accessories with copper wire and thin plastic and lead sheet.

The scratch-built details were based on photographs of the real thing and Tor admits that he was forced to estimate the scale, size and placement of some features. The end result is nevertheless quite impressive. Note the bracket on the hull rear which holds the exhaust muffler in place, the trackguards and rear driving light.

The finished model was painted with GSI CREOS Mr Color paints. The hard-edge camouflage pattern was achieved by masking off areas with Uhu-Tac and then spraying the next shade. The final touch was the extensive weathering and Tor used Acrylicos Vallejo and AMMO by Mig Jimenez products, including filters, washes and pigments, in addition to artist's oils.

JAGDPANZER 38 HETZER

1/35 SCALE

ERIC D WISDOM

Eric's replica is based on the Tamiya Mittlere Produktion Hetzer kit in 1/35 scale built with some photo-etched parts such as the machine-gun shield and the damaged fenders. Tamiya acrylics were used to paint the finished model and to achieve the camouflage pattern a base coat of Dark Green was first applied and then masked with Silly Putty. Red Brown was added and when dry the same process was used to apply Dark Yellow. This was followed by a wash of thinned Sepia artist's oil paint.

The crew figures are from Evolution Miniatures with one head coming from the spares box and other supplied by Alpine Miniatures.

The base is from Monroe Perdu Design Studios and was finished with arts and crafts acrylic paints, static grass and AK Interactive terrains snow.

JAGDPANZER 38 HETZER

1/35 SCALE

LIM KIAN GUAN

Singaporean modeller Lim Kian Guan, locally known as Chef Lim, is a frequent contributor to this series and I am more than happy to showcase his skill and creativeness once again. This model is based on the Tamiya 1/35 scale kit and was built almost straight from the box. The painting and weathering is extremely effective yet still subtle and the camouflage sheme is one of the factory-applied patterns that were employed during August and perhaps September 1944.

A rear view of the completed model. Interesting features to note here are the early exhaust muffler with its welded armoured guard, the toolbox typical of mid-production versions and the rear driving light.

Employing a similar camouflage scheme, Lim built this model using Dragon Models' 1/35 scale mid-production command tank. The Befehlspanzerwagen version of the Hetzer could easliy be distinguished by the large Sternantenna D in its armoured sleeve on the hull side.

Unlike many of the Dragon Models 1/35 scale kits this model comes with tracks made from individual links. It also contains some photo-etched parts and a turned aluminium barrel. Although dropped from the company's catalogue some time ago it can be found at certain retailers and on-line auction sites.

JAGDPANZER 38 HETZER

1/48 SCALE

RAMÓN SEGARRA

Ramón's model is based on the Tamiya Jagdpanzer Hetzer Mittlere Producktion kit in 1/48 scale, build straight from the box. With a length of just 131mm this kit includes an impressive amount of detail including the accurate exhaust cover, spare tracks and periscope guards.

Another view of the fully constructed model, less tracks, ready for painting and weathering.

Ramón chose one of the early camouflage schemes of a Dunkelgelb base with large irregular shapes, rendered in RLM Grun, outlined in Rotbraun. The dark yellow base colour was highlighted with a mixture of Dunkelgelb and white before the camouflage pattern was added. This stage was completed using Model Air Vallejo acrylics.

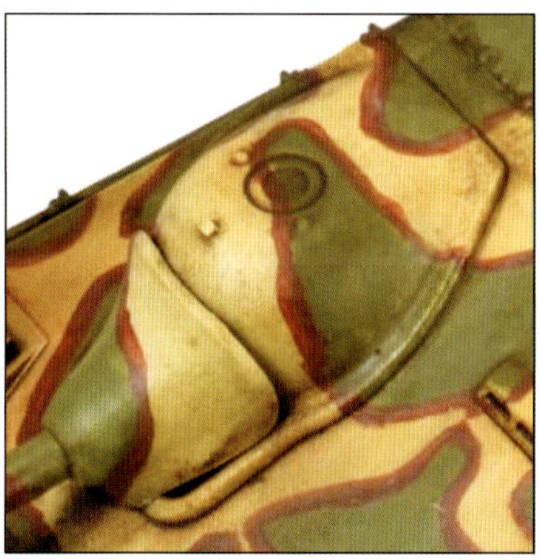

Small details such as the tools, machine gun and spare track links were painted by hand, also with Vallejo acrylics.

The completed model after painting and weathering. Here too Ramón made extensive use of products from Acrylicos Vallejo including a number of different washes and mud textures.

The suspension, wheels and tracks were weathered with Vallejo pigments including Natural Shadow, Sienna and Dark Steel 73.123.

Although the Jagdpanzer 38 Hetzer was in service for less than a year, it was an important addition to German army's arsenal and was a fundamental aspect of the programme to reorganise and reinforce the Wehrmacht's infantry formations. Indeed, the basic design concept of a lightly armoured, high speed tank killer built with a low silhouette and mounting a high-velocity gun, was to influence the development of self-propelled anti-tank systems up to the present day. It should come as no surprise that most plastic model companies have offered a kit of the Hetzer, of varying quality, in a number of scales.

As accurately as I can determine, the first kits were all released in 1972 by Italeri, Bandai, ESCI and Fujimi in 1/35, 1/48, 1/72 and 1/76 scales respectively. For their time these were all reasonably accurate kits but only the Italeri model is readily available today. In the early 1990s Dragon Models and Tamiya began producing kits of the Hetzer in 1/35 scale and these probably remain the most popular. The basic Dragon kit has been upgraded a number of times and in 2005 Tamiya released a highly-accurate 1/48 scale kit, the best offering in that scale. While these companies have received very little competition, even from the most recently-founded mainland Chinese firms, a large number of accessory sets have become available allowing unusual variations, of which there were many, to be built. As with the other books in the TankCraft series I have concentrated here on the popular scales of 1/35 and 1/48 although I have tried to include other sizes.

I should stress that the information presented on the following pages is far from definitive as many companies regularly withdraw products from their catalogues, often only to repackage and re-release them at some later date, and I would encourage readers to undertake their own research into the areas that interest them most. An index of manufacturers, together with suggestions for further reading, can be found on page 64 of this book.

Tamiya Inc. This Japanese company, the largest model manufacturer in the world, produces an extensive range of armoured vehicles, with complementary figures and accessories, and was almost solely responsible for the rise in popularity of 1/35 scale. Although rather basic by today's standards, the kits released from the early 1970s were highly accurate and detailed when compared to what else was available at the time. Importantly, they were easy to build and affordable and, like Airfix before them, Tamiya kits were all produced to a common scale which encouraged modellers to build a collection. The company's monopoly of the 1/35 scale market remained unchallenged until the release of the first Dragon Models kits in 1987. In 2003 Tamiya began releasing models in 1/48 scale and this has been extremely successful for the company, combining the potential for detail without the expense of the bigger kits. The current catalogue contains a mid-production Hetzer in 1/35 scale, which was upgraded with new parts as recently as 2012, and a 1/48 scale mid-production vehicle.

At left: Tamiya's 1/48 scale mid-production kit built and painted as suggested in the kit instructions. Below: Details of the mid-production 1/35 scale model.

Dragon Models. This Hong Kong-based company's early production Hetzer in 1/35 scale was originally released in 1994 and since then has been upgraded and repackaged several times, retaining the early production kit, and introducing a mid-production version and also a Befehlspanzerwagen 38. The same mid-production Hetzer kit, with the inclusion of a number of Waffen-SS infantry figures, was released under the company's Cyber-Hobby Orange Label range and this was still available at the time of writing. Parts from the basic model have also been utilised to produce a Bergepanzer 38 mit 2cm FlaK 38 and a 15cm s.IG.33/2(Sf) auf Jagdpanzer 38, both in 1/35 scale under the Smart Kit label.

The Hetzer kit released under the Girls und Panzer label, based on a popular Japanese television series, is the 1/35 scale Dragon Models mid-production vehicle.

The company's many small-scale kits are something of a benchmark and it is somewhat surprising that Dragon has never released a 1/72 scale version, either as one of their construction kits or as an assembled and pre-painted display model.

Above: Three Dragon Models releases based on the original 1/35 scale model. From left to right, the early production kit of 1994, the Orange Range mid-production model with figure set and the Befehlspanzer command tank, which was essentially the early production version with additional parts. Below: Dragon Models' 1/35 scale early production kit completed with the photo-etched brass detail set from Voyager Model. Note that the tubes fixed to the hull side at the centre, which hold the spare radio aerials, should be bent in the middle and not rendered in two separate pieces as shown here.

Below: Details of the Dragon Models' 1/35 scale early production kit including the drive sprocket with vinyl tracks and the Kugellaffette ball mounting and early gun mantlet. At far right is the 2cm Flak 38 kit built as a Bergepanzer recovery vehicle.

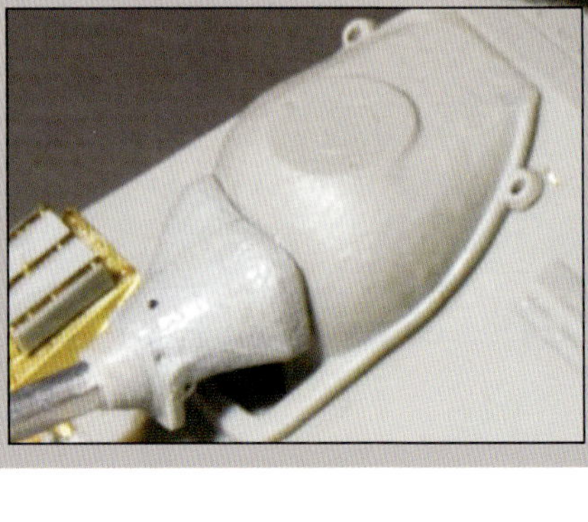

Academy Plastic Model. Based in Korea, this company's Military Miniatures range of 1/35 scale figures and vehicles includes a late production Jagdpanzer 38 which was first released in 2012 but has undergone at least two upgrades since then and offered with new parts as an early production model. Theses kits seem to be fairly accurate and provide options such as a selection of gun barrels and rear idlers. Crew figures are included with both kits. The same kits were recently re-boxed and sold under the Airfix label and by MPM in Poland.

Below: Academy's early production Jagdpanzer 38 Hetzer completed with the photo-etched upgrade set from Korean firm Artwox. Above: Details of the Artwox photo-etched parts applied to the Academy model.

Airfix and Revell. The recent 1/35 scale offerings from these companies are all re-packagings of older models, as explained under the headings for Academy Plastic Model and Italeri and I have included them here as a matter of interest only.

Italeri. This Italian company has been producing plastic models since its foundation in 1962 and the current catalogue contains a 1/35 scale Jagdpanzer 38 Hetzer which is based on the model that was first marketed in 1976 although some new parts have been included in the 2016 release. The 1/72 scale offering is the ESCI kit which was first produced in 1972 with new markings. On the other hand the 1/56 scale model is a completely new kit and while it is aimed at the wargames market, the detail is crisp and accurate and it can be built as a mid-production Hetzer, Flammpanzer 38 or Bergepanzer 38 mit 2cm FlaK 38. The 1/35 scale kit has also been released under the World of Tanks label. Revell's 1/35 scale Jagdpanzer 38 kit, released in 2018, is a re-boxing of the Italeri model.

Above: The latest box art for Italeri's 1/35 scale Hetzer. At right: The 1/56 scale Hetzer kit can be built as a standard tank destroyer or as the Flammpanzer and anti-aircraft tanks shown here. Although the suggested camouflage patterns are somewhat fanciful, the markings are accurate.

Sol Resin Factory. This Korean company, established in 1995, offers a huge range of complete kits, accessories and figures in 1/16, 1/35 and 1/48 scale. Shown below are details of the company's resin accessory set for a 1/35 scale late production Jagdpanzer 38 with the Flammvernichter exhaust muffler.

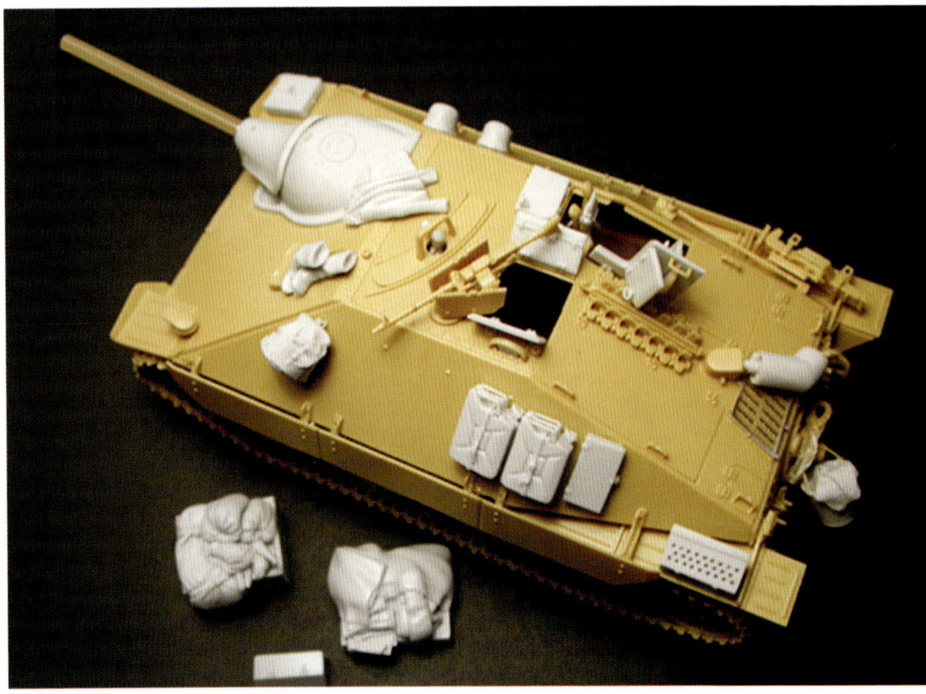

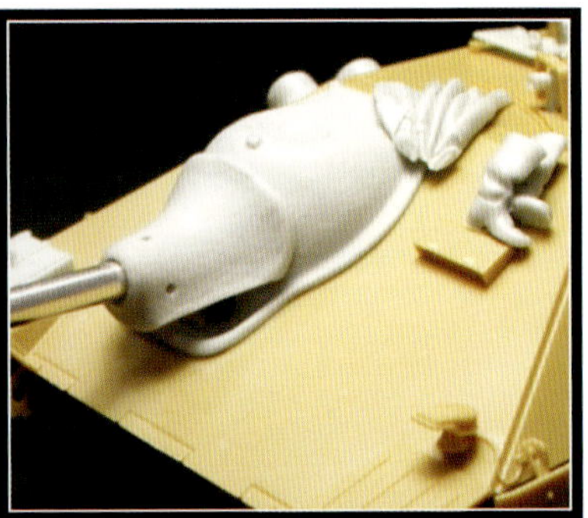

Blue Star Model Company. In 2012 this US-based company, which seems to have previously specialised in ship modelling, released a 1/16 scale Hetzer which could be built as either an early or mid-production vehicle. The kit contains 450 cleanly cast resin parts, an extensive array of photo-etched brass details and a set of markings printed to a very high quality. I have been unable to discover any further information on this company and they may no longer be trading, although there could be some connection to Chinese accessory producer Voyager Model. Confusingly, some kit boxes seem to be marked as 1/15 scale.

At left: Blue Star Model Company's 1/16 scale Hetzer built as an early production model. This kit was only ever available through the company's use of an on-line auction site and retailed for US$250 when it was first released. Hailed as the most accurate scale replica of the Jagdpanzer 38 at the time, the finished kit could be built as an early or late production model. There were at least two different boxes and one was indeed marked as 1/15 scale.

RB Model. RB Model from Poland produce a range of milled aluminium and brass gun barrels and resin accessories for 1/72, 1/48 and 1/35 scale armour models. Among other items the company produces highly detailed barrels for the 7.5cm Pak 38 L/48 gun in both 1/35 and 1/48 scale. There are also a large number of generic or universal items and shown below are the highly detailed towing shackles, radio antenna insulators and fuel containers.

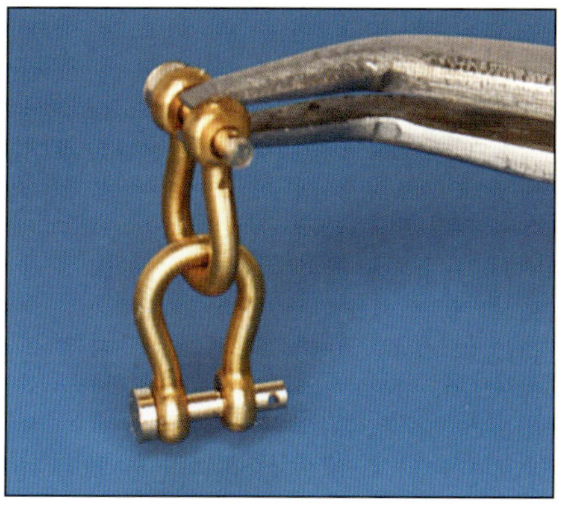

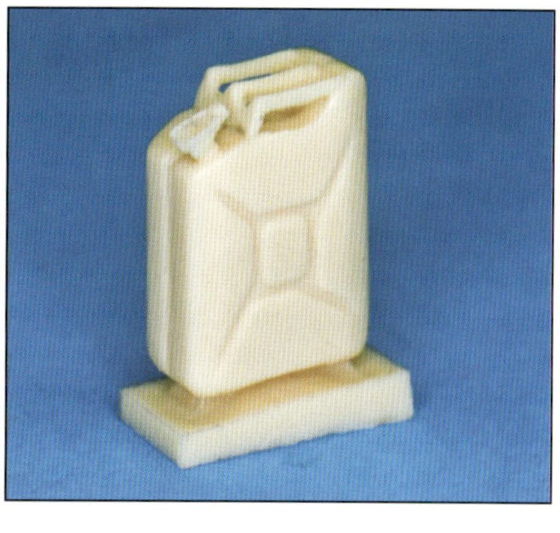

Eduard Model Accessories. Founded in 1989, this Czech company produces high quality upgrade kits in photo-etched metal and resin and a number of scale models. Currently available are extensive photo-etched detail sets to update the Academy and Dragon 1/35 scale models and both the 1/35 and 1/48 scale offerings from Tamiya. Eduard produced complete kits of the Hetzer from 2005 but unfortunately these highly detailed and accurate models have been discontinued.

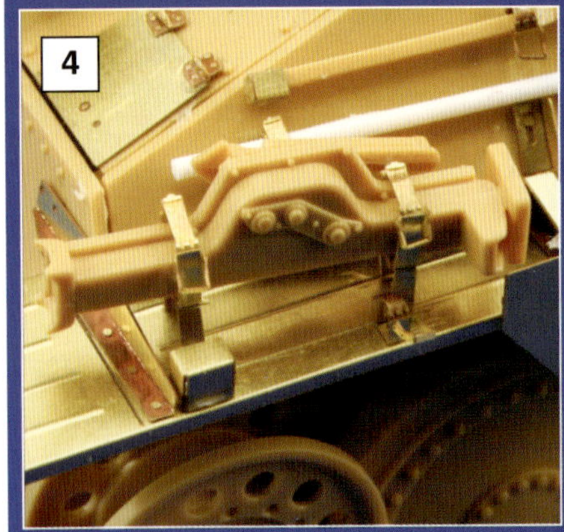

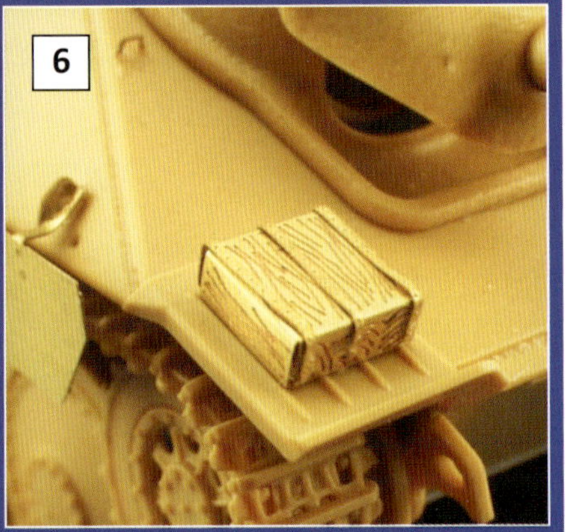

Above: 1 and 2. The detailed interior of Eduard's 1/35 scale late production Hetzer. 3 and 4. Photo-etched details for a late production Jagdpanzer 38. Although specifically marketed as being produced for the Tamiya kit, these parts could be used on any 1/35 scale model. 5 and 6. Details of the 1/48 scale upgrade set including the partly-perforated toolbox and rear driving light fixed to the trackguard brace.

Above: The Tamiya 1/35 scale Hetzer kit completed with Eduard's photo-etched brass and resin upgrade set. This also includes the crewman's belt buckle and uniform insignia.

Aber. This Polish company has been manufacturing upgrade sets since 1995, working in photo-etched brass, milled aluminium, stainless steel and even wood. For some time now Tamiya has included a number of Aber products with their models. A full list of the accessories made specifically for the Hetzer in all scales would be far too large to reproduce on this page, particularly in light of the company's regular new releases, and the reader can find the Aber's contact details on page 64.

Above: Tamiya's 1/35 scale Hetzer kit completed with Aber's photo-etched brass and aluminium detail set. Below: The Tamiya 1/48 scale mid-production model finished with Aber's upgrade set.

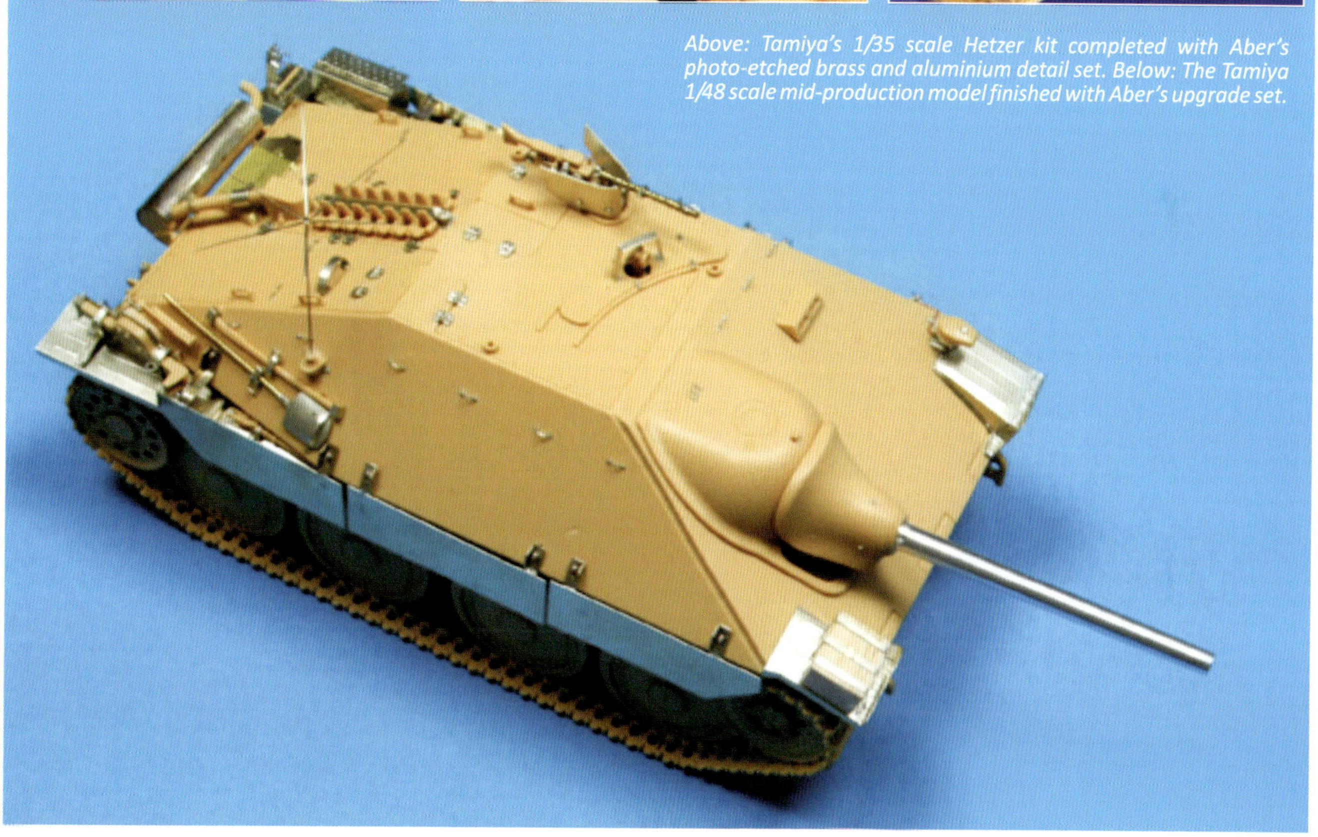

Griffon Model. This Chinese company produces a large range of photo-etched brass sets for the Hetzer 1/35 scale to fit most manufacturers' models, although they do recommend Tamiya and Dragon. Shown below, from left to right: Parts of the upgrade set for a mid-production model. Photo-etched metal and resin parts for a late production vehicles. A rifled barrel for the 7.5cm Pak 39 L/48 gun.

Hauler. Based in the Czech Republic, this company produces accessory sets in photo-etched brass and resin in 1/48, 1/35 and even 1/87 scale, a size which complements HO scale railway models. A number of items in the company's extensive range are now produced under the Bren Gun logo.

Below: Hauler's Flammpanzer conversion set in 1/48 scale produced specifically for the Tamiya kit. Inset: Details of the photo-etched brass upgrade set in 1/35 scale including a perforated cover for the exhaust muffler, air intake screens and early pattern toolbox.

Aftermarket Tracks. Most plastic kits today contain realistic tracks made from a form of vinyl which, unlike the older types, can be glued together, or hard plastic individual links. Nevertheless, there are many modellers who wish for the highest level of detail and realism possible and consequently the number of quality aftermarket products available seems to have increased in recent years, despite the extra expense and work required. Fully motorised models of course require workable tracks made up from individual links held together by pins.

Below: Examples of aftermarket tracks produced in various materials: 1. Early pattern tracks in white metal in 1/35 scale from Russian company Master Club. 2. Later type tracks, also in 1/35 scale from Master Club. 3. Early tracks in 1/72 sale made from resin from OKB Grigorov. 4. 1/35 scale injection moulded plastic workable tracks by Model Kasten from Japan. 5. Photo-etched brass early pattern tracks in 1/72 scale by Polish company Armory. 6. Individual metal track from Friulmodel.

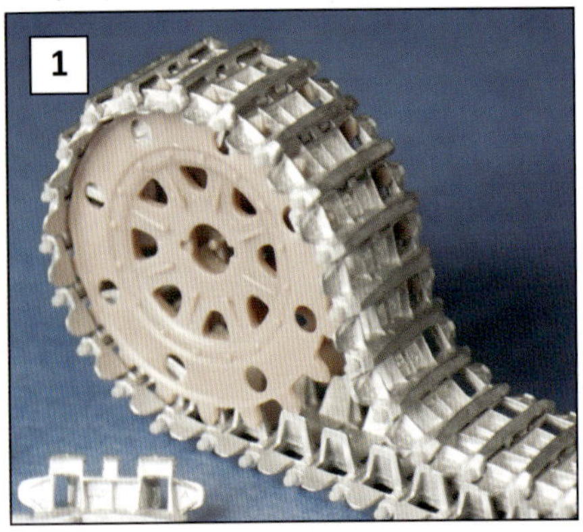

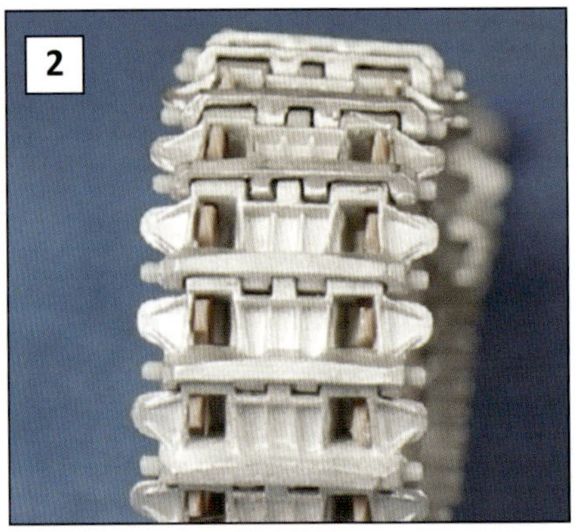

............text continued from page 16

The battalion arrived on the Western Front in the first weeks of April and took part in an attempt to recapture the village of Struth, west of Mühlhausen, in support of Kampfgruppe Worgitzki (1).

About half of the Hetzers were lost in this action. In late April the survivors found themselves encircled in the Harz Mountains and the commander, Major Lambert, ordered that all the remaining vehicles be destroyed.

During 1944 a series of independent anti-tank battalions were raised and equipped with the Jagdpanzer 38 Hetzer. Referred to as Heeres-Panzerjäger-Abteilungen, as they were to operate under army or army group command, they were numbered 731, 741, 743 and 744. All served in the East except the unit described below.

Heeres-Panzerjäger-Abteilung 741. This independent formation was formed on 25 July 1944 by rebuilding and renaming Armee-Panzerjäger-Abteilung 741. At this time the battalion contained two companies only as 2.Kompanie was detached and used to form Panzerjagd-Kompanie 1183 (2).

On Friday, 18 September 1944 the battalion was allocated fifteen Hetzers and a further fifteen were shipped from the HZA on the following Monday. On 25 September 1944 the battalion, probably the second and third companies and the headquarters element, was at the front under the command of II.Fallschirm-Korps and took part in the fighting for Cleve. This was almost certainly the first Hetzer-equipped unit to go into action on the Western Front.

On 3 October another fifteen Hetzers were dispatched and these were assigned to 1.Kompanie which very soon afterwards left for the East to act as part of an OKH reserve. Although I have not been able to determine when, or indeed if, this company returned, a report of 8 December 1944 states that thirty-nine Hetzers were on hand, just six vehicles short of the battalion's authorised allocation suggesting that all three companies were present (3).

The battalion later took part in the Ardennes Offensive, with just twenty operational tank destroyers, mostly notably in the fighting around Bastogne and the subsequent retreat through Vainden to Bitburg. In the spring of 1945 the battalion was engaged in the battles in the Reichswald and on the Wesel River bridgehead in support of a Kampfgruppe made up from 15.Panzergrenadier-Division and 116.Panzer-Division. By the end of March the battalion was retreating through north-western Germany and at that time twenty new Hetzers were assigned directly from Nachschub OB West.

Notes

1. Command by Oberst Hans-Heinrich Worgitzki, this Kampgruppe also contained Panzer-Einheit Dralle, which had been detached from Panzer-Lehr-Abteilung Bergen, with eight Sturmgeschütz III assault guns, one Pzkpfw IV tank and a single Panther. Heeres-Pionier-Brigade 688, which is often mentioned in connection to the unit, was probably at battalion strength or less.
2. This unit served in the East, at first with 83.Infanterie-Division and the from February 1945 until the end of the war with 6.Volksgrenadier-Division.
3. This also included three Bergepanzer 38 recovery vehicles.

Photographed east of Tangermünde on the Elbe River when units of 12.Armee surrendered to the US Army, this Hetzer was assembled at some time in early 1945 as evidenced by the triangular metal reinforcing brackets welded to the hull extensions and the large sheet metal cover over the recessed driver's visor. The camouflage pattern is typical of Skoda-built tank destroyers. The only units with 12.Armee at this time equipped with the Hetzer were Infanterie-Divisionen Potsdam, Scharnhorst and Ulrich von Hutten, all of which are located in the area on US Army situation maps, and Panzerjagd-Abteilung 3 .

This image was taken from the same series of photographs that includes the Hetzer shown on the previous page. The vehicle depicted here was probably built by BMM and has the large metal rings, one of which is visible to the right of the driver's visor, indicative of final production vehicles. Unfortunately, I have been unable to locate a side view of this Hetzer which would have better illustrated the rather bizarre face and eyes painted onto the gun mantlet or the name on the hull front.

Notes

1. The battalion was later equipped with four Sturmgeschütz III assault guns but this did not concern the company discussed here.

The last combat engagements involving Heeres-Panzerjäger-Abteilung 741 were probably fought between Wesermünde and Cuxhaven, on the North Sea coast, where the survivors surrendered to units of the British army. A vehicle of the unit is shown on page 17 of the Camouflage & Markings section of this book.

During the last months of the war several assault gun companies that had been attached to the Panzerjäger battalions of army divisions found themselves acting as independent units, usually because their parent formation had ceased to exist. In addition, a number of companies were formed with the intention that they would be attached to divisions but fought as independent units under army level command.

Records from this period are incomplete or missing entirely but it is known that the following units fell into these categories.

Jagdpanzer-Kompanie 1199. See Marine-Panzerjäger-Abteilung 2 below.

Jagdpanzer-Kompanie 1235. This independent unit was formed from personnel of Panzerjäger-Kompanie (Sturmgeschütz) 1235, the second company of Panzerjäger-Abteilung 235 from 198.Infanterie-Division. The division was badly mauled in the retreat through France and the company was detached in January 1945 and eventuaully sent to Truppenübungsplatz Milowitz, modern-day Milovice in the Czech Republic. One report states that Panzerjäger-Abteilung 235 had four Sturmgeschütz III assault guns on hand in the previous month and these were presumably handed over to Sturmgeschütz-Kompanie 2111 when it was attached to 198.Infanterie-Divsion in January, replacing the second company.

On 11 April 1945 a total of ten Hetzers and one Bergepanzer 38 recovery vehicle were shipped to the company but it is unclear if these actually arrived. At the end of the war the company was at Grafenwöhr in Germany and formed part of Kampfgruppe von Tolsdorff.

Jagdpanzer-Kompanie 1245. In July 1944 a Panzerjäger battalion made up of three companies was raised for 245.Infanterie-Division which was at that time stationed in northern France near Dieppe. The division was poorly equipped and the first record of any tracked vehicles being on hand with Panzerjäger-Abteilung 245 is from September 1944 when it was reported that a single 7.5cm Pak auf Lorraine (Marder II) self-propelled gun was available (1).

In late 1944 a company of Panzerjäger-Abteilung 245 was withdrawn from the front and sent to Milowitz to be equipped with Hetzer tank destroyers but the fourteen vehicles shipped to the company by the HZA on 7 December were diverted to another unit.

...........text continued on page 52

Just twenty Hetzers were converted by BMM to flamethrowing tanks and all were allocated to Panzer-Kompanien (Flamm) 352 and 353. Both vehicles shown on this page were disabled in the fighting for Gros-Réderching, south-east of Sarreguemines in France near the German border. The tank shown here, numbered S14, was towed to Oermingen were it was photographed and studied extensively by US Army intelligence officers. Unlike the standard Hetzer, these vehicles featured a hatch in front of the Rundumsfeuer shield and the handle is visible here. Note the hoop-shaped metal protective bracket for the hull periscope and the sighting mechanism on top of the Kugellaffette mount. Interestingly, the American report stated that a smoke ejector was fitted over the right rear fender but this is not possible to confirm from the available images. Note also the eight-hole rear idler with ribs welded in place.

Another Flammpanzer 38 of Panzer-Kompanie (Flamm) 353. This image provides a better view of the cover for the flamethrower which is damaged in the photograph above. Note the non-standard rack on the hull side which seems to have been intended to hold lengths of spare track. Note also that the hull extension towing eye appears to have been neatly cut off.

...........*text continued from page 50*

On 11 April 1945 a total of ten Hetzers and one Bergepanzer 38 were sent to the company, which was by this time at Grafenwöhr in Germany, and Jagdpanzer-Kompanie 1245 may have seen some action with Kampfgruppe von Tolsdorff in the last days of the war.

Jagdpanzer-Kompanie 1257. In June 1944, 2.Kompanie of Panzerjäger-Abteilung 257 was detached from its parent formation and sent to Milowitz in Czechoslovakia to be retrained and was re-equipped with fourteen Hetzer tank destroyers.

In August, before the company could rejoin the battalion, the other two companies were almost completely destroyed in the fighting on the Eastern Front. The second company, now referred to as Sturmgeschütz-Abteilung 1257, returned to the East at the end of August and served as an independent unit under Heeresgruppe Süd. In early November 1944 the remaining personnel were withdraw to Milowitz with the intention that the company would form part of the Panzerjäger battalion of the newly-formed 257.Volksgrenadier-Division.

In January 1945 a consignment of fourteen Hetzers and one Bergepanzer 38 intended for Sturmgeschütz-Abteilung 1257 were diverted to another unit and the company, by now renamed Jagdpanzer-Kompanie 1257, was transferred to Grafenwöhr in Germany. On 11 April 1945 ten Hetzers and one Bergepanzer 38 were dispatched to the company by the Heereszeugamt and Jagdpanzer-Kompanie 1257 is last mentioned as being in combat in the Grafenwöhr area as part of Kampfgruppe von Tolsdorff, fighting against US Army troops who captured the town on 21 April 1945. The company commander, Leutnant Horst Sieber, was awarded the Ritterkreuz for the bravery and leadership he displayed during these actions. See also 257.Volksgrenadier-Division above.

Panzerjäger-Kompanie zbV AOK 1. This unit, nominally under the control of Armeeoberkommando 1, was attached to 198.Infanterie-Division in late 1944 and may have been raised to help rebuild that formation. Records show that the company was allocated fourteen Hetzers on 19 January 1945 but it appears that just one of these vehicles was on hand at the end of the month.

Panzerjäger-Kompanie Bock. Formed in late 1944, Panzerjäger-Kompanien Bock, Lang and Pankow were each allocated fourteen Hetzers which were shipped from the HZA on 26 and 27 November. These companies all lacked a staff element and repair sections and for a time it was planned that they should form the Panzerjäger component of the Führer-Begleit-Brigade.

Photographed in Germany in 1945, this Flammpanzer 38 must be from Panzer-Kompanie (Flamm) 352, the only unit equipped with these tanks after the battles of Operation Nordwind. Just twenty Hetzers were converted to the flamethrowing role in early December 1944 and all had been allocated by the end of the same month. Interestingly, this vehicle has the large rings welded to the side of the hull, clearly visible here above the Balkenkreuz, which were thought to have been introduced into production from early 1945 in the late production phase. This Flammpanzer is also shown in the Camouflage & Markings section on page 20.

A front view of the Flammpanzer number S-14 depicted on page 51. Note the heavy reinforcing plate welded to the front of the Kugellafette gun mount. The sheet metal cover for the flamethrower was easily damaged as can be seen here. It was designed to disguise the weapon's true purpose and Allied units initially though these vehicles were a howitzer-armed model of the standard Hetzer.

All three companies were eventually subordinated to Heeresgruppe G and took part in Operation Nordwind, supporting the divisions of 1.Armee. In late January 1945 the remaining vehicles and personnel were amalgamated with the remnants of Panzerjäger-Kompanie 708 to form Panzerjäger-Abteilung zbV 'G' (1). It is possible that the remnants of this battalion were attached to II.Abteilung, Panzer-Regiment 2 as part of Kampfgruppe von Luttichau.

Panzerjäger-Kompanie Lang. See Panzerjäger-Kompanie Bock above.

Panzerjäger-Kompanie Pankow. See Panzerjäger-Kompanie Bock above.

Panzerjäger-Kompanie Gekeler. Formed in early 1945, the company was allocated ten Hetzers on 11 January and took part in Operation Nordwind as part of Heeresgruppe Oberrhein. This unit is not, as suggested in at least one source, Panzerjäger-Abteilung zbV 'G'.

In December 1944, acting on a suggestion made by Hitler, it was decided to convert a number of Jagdpanzer 38 Hetzers to a flamethrowing role. Although these were obviously not tank destroyers, and therefore outside the limits of this study, some mention should be made of these unusual vehicles. Fitted with the Koebe-Gerät Flammenwerfer 41 in place of the 7.5cm main gun, a total of twenty vehicles were ready within weeks and allocated to the companies listed here.

Panzer-Kompanie (Flamm) 352. Formed in December 1944, the company was allocated ten Flammpanzer 38 in the same month and arrived at the front on Christmas Day. Panzer-Kompanie (Flamm) 352 took part in Operation Nordwind and supported elements of 25.Panzergrenadier-Division in the attacks on the Maginot Line defences in the first week of January 1945 near Oberroedern, north-east of Hagenau. Later that same month the company was engaged in the final attempt to take Rittershoffen which one German commander described as: 'One of the hardest and most costly battles that ever raged'.

In early February the surviving vehicles and personnel of Panzer-Kompanie (Flamm) 353 joined the company and by 15 March 1945, Panzer-Kompanie (Flamm) 352 reported that nine Flammpanzer 38 were on hand, of which one was in repair. From this it would seem safe to assume that four vehicles were lost in action up to that time.

Panzer-Kompanie (Flamm) 353. Formed in December 1944, the company was equipped with ten Flammpanzer 38 tanks and had arrived at the front by the end of the month.

On 9 January 1945 the company took part in an assault on the village of Hatten in support of 17. SS-Panzergrenadier-Division with disastrous results, losing seven Flammpanzer, and by early February the remnants of the company were absorbed by Panzer-Kompanie (Flamm) 352. It should be mentioned that in some sources, and official documents, these short-lived units are referred to as Panzer-Flamm-Kompanien.

Schwere Panzer-Abteilung 507. While refitting at Truppenübungsplatz Milowitz this heavy tank battalion was allocated ten Hetzer tank destroyers. The crews of 3.Kompanie used these to fight their way to the West through Punzlau, Lissa and finally Rosenthal where they surrendered to US forces (2).

Notes

1. The German term zur besonderen Verwendung, usually abbreviated to zbV, can be translated as meaning 'for special purpose'.
2. The history and organisation of this heavy tank battalion is examined in some detail in the first book in this series *Tiger I and Tiger II: Tanks of the German Army and Waffen-SS, Eastern Front 1944* and *TankCraft 13 Tiger I and Tiger II Tanks: German Army and Waffen-SS, The Last Battles in the West.*

Jagdpanzer 38 Hetzers were allocated to formations of the Waffen-SS from September 1944 but most units served in the East. Those that are known to have taken part in the fighting on the Western Front are listed below.

SS Panzerjäger-Abteilung 38. Attached to 38.SS-Panzergrenadier Division Nibelungen, the battalion's 1.Kompanie was equipped with ten Jagdpanzer 38 Hetzers which were shipped on 15 April 1945 and one Bergepanzer recovery vehicle dispatched on the following day. The second company was outfitted with towed 7.5cm Pak 40 anti-tank guns and the third company was a Flak unit. Most of the battalion's personnel were drafted from the Waffen-SS Prinz Eugen and Nord divisions.

The battalion was temporarily subordinated to 17.SS-Panzergrenadier-Division Götz von Berlichingen until 24 April when it returned to its parent formation. Interestingly, a report from that time shows that a single Hetzer was on hand with SS-Panzer-Abteilung 17 (1). As part of XIII.SS-Armeekorps the division fought against units of the US Army along the Danube between Vohlburg and Kelheim but by 1 May 1945 had been pushed back as far as Wasseburg, about 40 kilometres east of Munich. The last two serviceable Hetzers were lost in a firefight in the streets of Dorfen, just north of Wasseburg, at this time. The remnants of the division, which had only ever reached the strength of a strong brigade, surrendered on 8 May 1945.

Begleit-Bataillon Reichsführer-SS. Confusingly, several combat units were raised during the war with this title and this formation should not be confused with the battalion which ended the war fighting in Berlin. Probably raised from elements of SS-Panzerjäger-Abteilung 6 in December 1944 or January 1945, when Himmler was appointed to command Heeresgruppe Oberrhein, Begleit-Bataillon Reichsführer-SS was subordinated directly to 19.Armee and was reportedly made up of two Panzerjäger companies. The battalion is said to have taken part in the unsuccessful attempt to recapture Strasbourg, as part of Operation Nordwind, but the first recorded allocation of ten Jagdpanzer 38 Hetzers did not take place until 25 January 1945, the very day a halt was called to the offensive (2).

These vehicles were reported as still being with the battalion in March. On 9 April 1945 the remaining personnel and equipment of Begleit-Bataillon Reichsführer-SS were permanently attached to 38.SS-Panzergrenadier Division.

In the last months of the war four divisions and two brigades were formed from men of the Reichsarbeitsdienst (RAD), Germany's national labour service. Initially referred to as RAD-Divisionen 1, 2 and 3, the formations discussed here were renamed Infanterie-Divisionen Schlageter, Friedrich Ludwig Jahn and Theodor Körner respectively in April 1945. Efforts to integrate the RAD units into the Volkssturm were strenuously resisted by Reichsleiter Konstantin Hierl, the head of the RAD, and for the most part the units retained their former ranks and uniforms (3). Although none actually received any Hetzer tank destroyers, allocations were made by the HZA and I have included them here for the sake of completeness.

Panzerjäger-Abteilung Schlageter. Attached to Infanterie-Division Schlageter, formerly RAD 1, the battalion's 4.Kompanie was allocated ten Jagdpanzer 38 tank destroyers on 17 April 1945 but these were never received and went instead to an ad-hoc unit formed around Stab, Panzerjäger-Abteilung 553.

The division first went into action near Ludwigslust, south-east of Hamburg, later in the same month in support of the divisions of 3.Armee. The survivors of the division surrendered on 3 May 1945. See Panzerjäger-Abteilung 1553 above.

Panzerjäger-Abteilung Friedrich Ludwig Jahn. Attached to Infanterie-Division Friedrich Ludwig Jahn, the Panzerjäger battalion's 4.Kompanie was allocated ten Jagdpanzer 38 tank destroyers on 16 April 1945 but all were diverted to an ad-hoc unit formed around Stab, Panzerjäger-Abteilung 553. See Panzerjäger-Abteilung 1553 above for further explanation.

Panzerjäger-Abteilung Theodor Körner. Although it was proposed to build the battalion around Jagdpanzer-Kompanie 1159, which was training at Milowitz and was in the process of receiving ten Hetzers, the company was used to form part of Gemischte, or mixed, battalion formed around the headquarters element of Panzerjäger-Abteilung 553. See Panzerjäger-Abteilung 1553 above.

Marine-Panzerjäger-Abteilung 2. The battalion was part of 2.Marine-Infanterie-Division raised in February 1945, largely from sailors of the Kriegsmarine, Waffen-SS replacement units and boys of the Hitler Jugend. The ten Hetzers promised to the battalion were never dispatched and in their place the personnel of Jagdpanzer-Kompanie 1199, an army unit, were attached to the division.

But by mid-April the company had not received any vehicles and it is possible that the men remained at the Milowitz training area in Czechoslovakia until the end of the war. At least one source suggests that the Jagdpanzers intended for this company were misappropriated by another unit which was training at Milowitz. The battalion's other companies had no heavy equipment and were deployed with bicycles and armed with Panzerfausts.

Notes

1. This cannot be easily explained by the suggestion, probably correct, that the Panzerjäger battalion used the division's repair and maintenance facilities as the same report shows the battalion's full allocation of ten as being on hand.
2. There is also a record of ten Hetzers on hand with Heeresgruppe Oberrhein in January 1945 and some connection cannot be ruled out.
3. This did not, of course, extend to the large numbers of Wehrmacht personnel who were attached to these divisions.

Like most weapons systems the Hetzer underwent numerous modifications during its production life. Indeed, for a vehicle that was manufactured for just under twelve months a bewildering number of changes, usually minor, were introduced. For example, the suspension alone accounted for four changes to the roadwheels, nine distinct rear idler designs and the use of five different track types. The production runs discussed below are divided here into early, mid, late and final periods and I should stress that these terms are post-war inventions and used here as a matter of convenience only. Surviving records are unfortunately incomplete, particularly for the last months of the war, and some of the information presented here is based on photographic evidence. The reader should also bear in mind that the revisions and modifications listed below were introduced as new parts or equipment became available and it is common to see features defined as early or mid-production on vehicles manufactured in the spring of 1945. Examples of this are shown in the Camouflage & Markings section which begins on page 17 and also in the photographs reproduced throughout this book.

Those who are new to this series may not be familiar with some German terms and a brief explanation may be helpful here. The Heereswaffenamt (HWA) was the German army's ordnance office. It was responsible for the procurement of all equipment, including weapons and ammunition, and also research and development. The Heeres-Abnahmestelle as mentioned in this narrative was a branch of the Heereswaffenamt responsible for the acceptance of equipment. The specific office referred to in the text, Heeres-Abnahmestelle II(K), was responsible for the Hannover area. The Heereszeugamt (HZA) was the army's supply directorate and vehicles and equipment were shipped to units through its various offices, or Heereszeugämter, in each military district.

1. April and May 1944. The Early Production Model.

Series production of the Hetzer began at Böhmische Märische Maschinenfabrik (BMM) in April with a target of twenty fully assembled vehicles proposed for the first month.

These early examples retained the 32-bolt roadwheels and the 12-hole flat rear idlers of the prototypes. The practice of drilling eight holes in the outer ring of the drive sprocket was discontinued in an effort to reduce production time and an armoured cover for the final drive housing was introduced. The 293mm-wide Pzkpfw 38t track of the prototypes was replaced by a purpose-built 350mm-wide Hetzer track. This early pattern track may have been fitted to some production vehicles as late as September 1944.The distinctive ram's horn front towing brackets bolted to the lower front plate of the prototypes were replaced by extending the hull sides, at both front and rear, and drilling holes to create simple towing 'eyes'. Small brackets were welded to the hull sides and front to hold a tarpaulin.

The stepped gun mantlet of the prototype was fitted to the first twenty production vehicles (Fgst. Nr. 321004 to 321023) but the sides were first machined flat. A lighter version with smooth edges was introduced and fitted to the fifty Hetzers assembled in May (Fgst. Nr. 321024 to 321073).

The Type I Kugellaffette, the ball-mount which held the 7.5cm Pak 39 gun, was fitted to the prototypes and the first few production vehicles. This was held in place by three external bolts on the right side and four on the left. During the early production run this was replaced by the simpler Type II or Type III, both of which

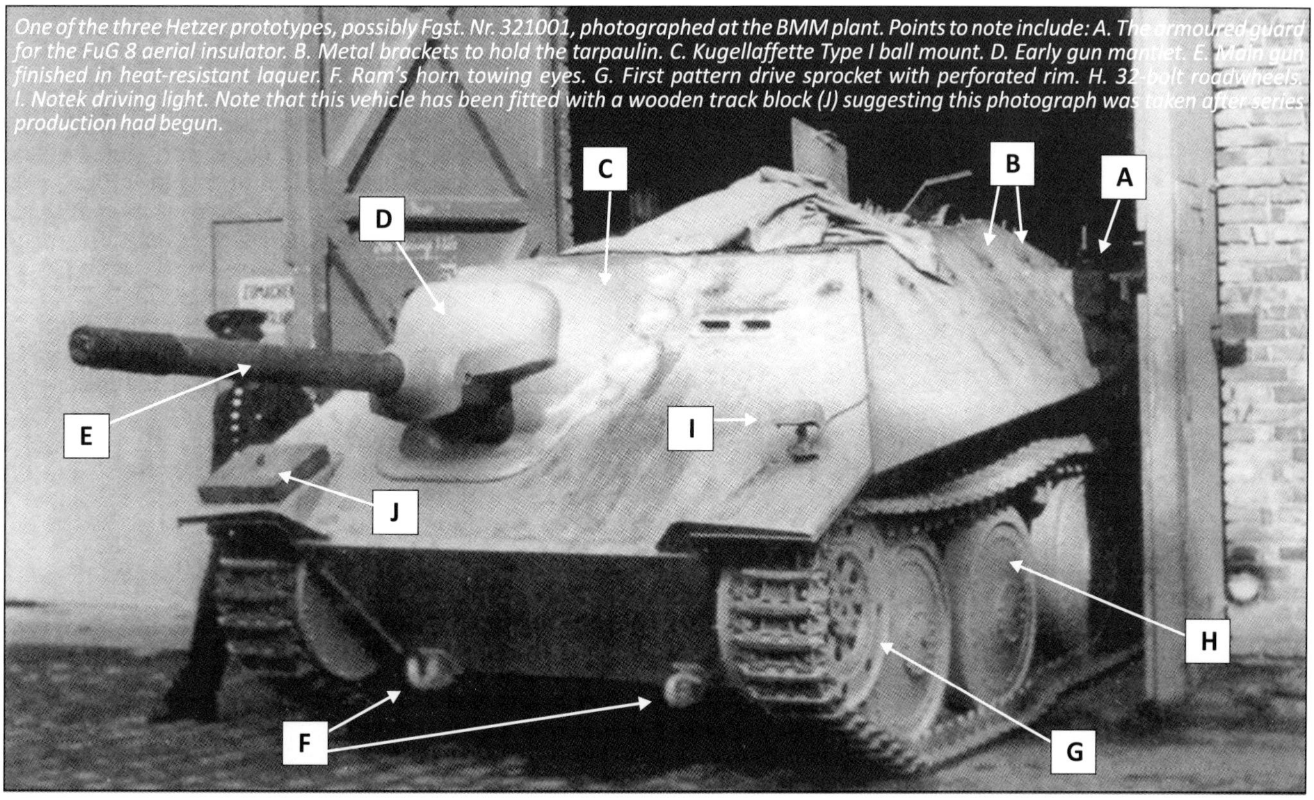

One of the three Hetzer prototypes, possibly Fgst. Nr. 321001, photographed at the BMM plant. Points to note include: A. The armoured guard for the FuG 8 aerial insulator. B. Metal brackets to hold the tarpaulin. C. Kugellaffette Type I ball mount. D. Early gun mantlet. E. Main gun finished in heat-resistant laquer. F. Ram's horn towing eyes. G. First pattern drive sprocket with perforated rim. H. 32-bolt roadwheels. I. Notek driving light. Note that this vehicle has been fitted with a wooden track block (J) suggesting this photograph was taken after series production had begun.

A series production Jagdpanzer 38 photographed in early 1944 at the BMM plant in Czechoslovakia. This vehicle has a number of early production features including: A. Large engine access door. Note the lack of filler hatches which would be incorporated in the mid-production series. B. Early pattern exhaust muffler with perforated cover. C. Rear light fixed to trackguard brace. D. Perforated toolbox. E. Hull extension towing eye. F. Spare track links. The upper set would later be replaced by a filler hatch. G. First pattern rear idler. H. Hull Schürzen with brackets welded in place and without the inward bend at front and rear.

were fastened to the hull glacis by two external bolts. The Kugellaffette IV, which was very similar to the latter two, may have been introduced late in this production phase. The SflZF 1a gunner's sight, with its distinctive square top, was fitted to the early-production and possibly most of the mid-production vehicles. The pistol port of the prototype was deleted on the production series vehicles.

The exhaust mufflers of the three prototypes and the first twenty vehicles assembled in April were fitted with a perforated cover. At the centre of the engine deck a cast armoured guard covered the exhaust pipe where it entered the hull. While most of these guards had rounded edges, some were flat.

Hull Schürzen were introduced on the production vehicles and were initially attached by brackets bolted to the hull and welded to the outside face of the Schürzen. Photographs of Fgst. Nr. 321003 taken in 1945 suggest that these armoured skirts were probably retro-fitted to the prototypes.

The wires for the Notek driving light, which entered the hull through the roof on the prototypes, now led to a small hole on the left hand side.

The armoured driver's visor housing was retained but a revised, shortened shield for the Rundumsfeuer machine gun was introduced. Photographic evidence shows that due to a shortage of these guns a number of vehicles were assembled with a metal plate covering the mounting. The perforated stowage box fitted to the left rear trackguard of the prototypes was retained and a convoy light was fitted to the trackguard brace behind the box on production vehicles. Two bent aerial tubes were fitted to the hull on the left side. On the right side the position of the jack was reversed and a jack block was carried on the right front fender.

The hull nose plate was now interlocked unlike the straight welded join of the prototypes. A tow shackle assembly was bolted to the rear plate and three spare track links were held on the lower left side. Another three links were carried on the engine deck on the right. In May the large hatches on the rear deck were replaced by a hatch for the commander which opened to the rear, and separate smaller hatches for the radiator on the right and fuel tanks on the left. Three sockets, or Pilze, were welded to the roof to accommodate the 2-ton jib boom. It is possible that very few vehicles were fitted with Pilze in May and this modification may not have become standard until June.

The rectangular armoured cover for the FuG8 radio antenna mount and insulator that had been fitted to the prototypes was replaced by a curved version.

In May the first Bergepanzer 38 recovery vehicles were assembled and these were built side by side with the Hetzer and were

...........*text continued on page 58*

A

B

C

...........text continued from page 56

Notes

1. Fgst.Nr. 321042, on display today at Camp Borden in the United States, has many of its original features including this towing bar.

2. The RAL colour system and its use by the German armoured troops is explained in some detail in the other books in this series including *Tiger I & Tiger II Tanks: German Army and Waffen-SS, The Last Battles in the West 1945* and *Panther Tanks: German Army and Waffen-SS, Defence of the West, 1945.*

not allocated their own Fgst.Nr series. In June the brackets for carrying the unditching beam, the jib boom crane arms and the rigid towing bar on the hull rear plate were added. This latter modification was sometimes added to the Jagdpanzer version (1).

All vehicles were painted in a base coat of Dunkelgelb RAL 7028 and units were authorised to apply camouflage patterns using Olivgrün RAL 6003 and Rotbraun 7028. The two latter colours were supplied in 2kg tins of paste which could be thinned with either petrol or water (2). The use of Dunkelgelb as a base coat at this time is quite possibly the only aspect of Hetzer camouflage that is not the subject of debate.

The Heereswaffenamt (HWA) accepted all twenty production vehicles completed by BMM during April as well as the three prototypes (Fgst. Nr. 321001 to 321003). The planned production figure of fifty for May was also achieved and all were accepted by the HWA. At the end of May 1944, fifteen of these vehicles were allocated to testing facilities including Hillersleben, Kummersdorf, Bergen and the gunnery school at Putlos. A further seven were shipped to the Panzerjäger-Schule at Mielau in Poland and photographs show that least one of these was a Befehlspanzer command vehicle, almost certainly one of the prototypes.

2. June to September 1944. The Mid Production Model.

It was soon realised that the proposed production figures for the Hetzer would not be achieved by a single company and the firm of Skoda, BMM's long-time rival, was awarded a contract to build 2,000 vehicles beginning with Fgst. Nr. 323001. The first completed vehicles left the Skoda assembly lines in July and during the mid-production period the following modifications were incorporated by both companies.

The 32-bolt roadwheels were enlarged by fitting them with a wider rim which was initially bolted to the outside edge of the centre disc but later simply riveted (see below). In August a new roadwheel with a larger-diameter centre disc and a narrow rim was introduced and, as an economy

An early production Hetzer photographed in 1945, probably in Czechoslovakia. Points to note are. A. The single bolt on the left side of the Kugellaffette II or III. An identical bolt held the mount in place on the right side. B. The early pattern gun mantlet with machined sides. C. The wooden jack block. D. The hull extension towing eye. E. The Notek driving light. Note that the lead now enters the hull on the right side. F. The later style of armoured cover for the FuG 8 aerial insulator. Note that is now rounded and not angular as those fitted to the prototypes were. G. The cast driver's visor. This was replaced by a recessed version in the late production series. H. The gunner's sight. This is the SflZF 1c model, distinguished by its rounded top, fitted to late production vehicles.

Previous page: Examples of the factory-applied Ambush camouflage schemes introduced in August 1944. These are also described in the text. A. The basic camouflage pattern used by BMM was made up of meandering cloud-like patches of Olivgrün RAL 6003 and Rotbraun RAL 8017 on a base coat of Dunkelgelb RAL 7028. The official orders called for the darker colours to cover two-thirds of the surface but the ratio seems to be much higher here. Over the patches, small chips and wedges of contrasting colour have been applied. These vary in shape but a pattern is discernible and they were certainly applied to this vehicle using a stencil. B. An example of the Ambush camouflage scheme applied to a mid-production Jagdpanzer 38. Note that the contrast between the Olivgrün and Rotbraun is very low and the edges are slightly feathered, contrary to official instructions. The chips of contrasting colour here seem to have been applied quite haphazardly. C. Built in August or September 1944, this Hetzer is finished in the Ambush camouflage scheme typical of Skoda-built vehicles. The two darker shades have been painted in irregular shaped patches over the Dunkelgelb base. The chips of contrasting colour varying in shape and size and may have been applied individually with a brush. A further example of the Ambush scheme is shown on page 17 of the Camouflage & Markings section. This type of camouflage may have only been in use during the last two weeks of August and the first half of September suggesting that it was applied to approximately 200 vehicles with as few as thirty of that number being built by Skoda.

measure, the rim was increasingly drilled to accommodate sixteen bolts only. By the end of August a series of different rear idlers had been introduced with the 12-hole flat version initially replaced by a 6-hole model although it seems that Skoda continued to utilise the former. At least some BMM-assembled vehicles were fitted with an 8-hole rear idler. The drive sprocket and armoured final drive housing were unchanged. The early pattern track was slowly replaced by a version which was cast with a reinforced centre section.

Probably beginning in August, a new gun mantlet with smooth edges was introduced and the Type V Kugellaffette, without external bolts, entered production. This modification reduced the overall weight by 200 kilograms. The early-pattern exhaust and cast armoured guard were retained. The driver's visor housing and the shield of the Rundumsfeuer machine gun were unchanged.

As with the early production models, the hull Schürzen were made up of three sections on both sides and each section was held in place by two brackets which were bolted to the hull. From the beginning of June the brackets were fixed to the inside face of each section and held in place by two external rivets. Towards the end of this production run the corners of the forward and rear sections were bent inwards. The two bent aerial tubes on the hull left side were moved towards the front and the perforated stowage box was gradually replaced by a solid-sided version. The convoy light on its post was fitted to the trackguard support behind the toolbox on the first production models but the light, without the post, was later attached directly to the trackguard brace.

The position of the jack and jack block was unaltered. The earliest mid-production Hetzers had no filler hatch on the rear engine deck while those assembled later in this production run had one on the right and later one on each side.

On 19 August 1944 the HWA advised assembly plants that the practice of units in the field creating their own patterns of camouflage would halt immediately and that a more-or-less standard scheme would be applied to vehicles before they left the factories. The base colour of Dunkelgelb RAL 7028 was to be retained, over which large patches of Olivgrün RAL 6003 and Rotbraun RAL 8017 were to be painted. Many firms took this a step further and added contrasting dots or wedges to the patches of colour. This is commonly referred to today as the Ambush Scheme or Hinterhalt-Tarnung and although the latter may sound official there is no evidence that either term was used during the war. The pattern applied to Hetzers was made up of cloud-like patches of the two darker colours, covering more than two-thirds of the surface area. Small triangle or wedge shapes of contrasting colour were applied to BMM-assembled vehicles while the camouflage of Skoda-produced Hetzers was finished with dots or rectangular chips. At both factories the wheels were painted in all three colours, predominantly the darker shades, but the Ambush camouflage was seldom, if ever, extended to the barrel of the main gun. This was most often painted in a combination of the two dark shades or simply left in its coat of very dark, almost black, heat-resistant lacquer. This camouflage scheme was short-lived and may have only been applied to the vehicles which left the assembly lines in the latter half of August and early September.

BMM was able to complete 100 vehicles in June and all were accepted by the HWA. In July, a further 100, of a planned 175, left the company's Brünn factory. In the same month production began at Skoda, as mentioned above, with just ten Hetzers completed. With assembly now in full swing a total of 123 vehicles were allocated to operational units during July with the first going to the independent Panzerjäger-Abteilungen 731 and 743 which each received forty-five, their full complement. Officially at least this number would have included ten Befehlspanzer command vehicles. A plan to send fifteen Hetzers to Romania was abandoned. In August 1944 BMM produced 150 vehicles while the Skoda plant completed twenty. In that month a total of 124 vehicles were shipped by the Heereszeugamt (HZA) with the majority going to the Panzerjäger battalions of infantry divisions.

3. September to December 1944. The Late Production Model

The drive sprocket and final drive housing were unchanged but the 16-bolt roadwheels became standard. From October the rims of roadwheels were secured by rivets. A mixture of rear idlers were incorporated into production including a 6-hole version, an 8-hole model and an 8-hole wheel with ribs welded in place. It is possible that the latter were only fitted to BMM-assembled vehicles.

In December the lower edge of the hull extensions which formed towing eyes was straightened. Examples of the new design are shown in the Camouflage & Markings section. The cast armoured cover for the driver's periscope was dropped from production and a recessed version with sheet metal cover to keep out rain was introduced. The locking bar and handle of loader's hatch were reversed and the rear portion of the commander's hatch was now made from two pieces. The earlier gunner's sight was replaced by the SflZF 1c model.

The rear engine deck was redesigned with a small filler hatch on either side. The right

A Jagdpanzer 38 Hetzer of Panzerjäger-Abteilung 1600 photographed in western Czechoslovakia in April 1945 just before this unit attempted to surrender to the US Army. Note the large sheet metal cover over the driver's visor, a feature of final production models. Note also the the right side towing eye has been broken, a common sight. The Fgst. Nr., which is unfortunately indecipherable, is just visible on the hull glacis. The letter A on the lower edge of the gun mantlet may indicate a platoon or battery but may also have some connection to the recoil mechanism lubricant and is also seen, rendered in the same style, on Pzkpfw V Panther tanks at this time.

side hatch handles were vertical while those on the left were horizontal on Hetzers built by Skoda. Both handles are placed horizontally on BMM-assembled Hetzers. The cast cover for the exhaust tube was replaced by a square welded version. The horizontal muffler of the early and mid-production models was prone to backfiring and overheating to the extent that the glow could be seen at night and from October a Flammvernichter exhaust muffler, set at a 45-degree angle, was incorporated into production. The earlier muffler was never completely replaced, however, and it can be seen fitted to vehicles built in 1945.

A new track with four reinforcing ribs was introduced in December. At about this time the brackets which supported the hull Schürzen were welded to the inside face of the skirt. A boarding step was welded to the blowtorch cover on the rear hull and the left side spare tracks were positioned lower while the right side links were moved slightly higher. A number of other modifications were incorporated into the late production version but were mostly internal, such as the rearrangement of the 7.5cm ammunition stowage and provision of crew heating, and I have not included them in detail here.

Towards the end of October 1944, manufacturers were advised that the use of Dunkelgelb RAL 7028 as a base colour for armoured vehicles was to be discontinued. Tanks were from now on to be painted in a camouflage scheme made up of patches of Braun, Grün and Gelb over the basic red oxide primer (1). This order may have in fact been the formalisation of a practice that had begun at some factories in September at around the time Zimmerit was discontinued (2).

At BMM this initially took the form of wide bands of the three colours applied with hard edges but gradually the band of Dunkelgelb became quite narrow and meandering and the two darker shades predominated.

In a number of photographs the contrast between the colours in this latter pattern is quite low, suggesting the paint was thinned. The Skoda camouflage was almost identical to that employed by BMM but the wide bands of colour were retained until the end of production. Matte black rectangles were painted on the hull front plate as decoys to draw the enemy's fire away from the driver's visor.

On 29 November 1944 the HWA advised vehicle manufacturers that a new camouflage system was to be introduced where the armour components were to be painted in a base coat of Dunkelgrün RAL 6003 before they were delivered to the assembly plants. Once assembly was completed, the tank would be camouflaged by the addition of Rotbraun and Dunkelgelb painted in sharp contours. This scheme was not to come into effect until March 1945 but on 20 December 1944 a new order directed that the painting of armour components in Dunkelgrün was to begin immediately. Just how widespread this practice became is uncertain and it is almost impossible, in most cases, to differentiate between a

Notes

1. Although these names are not strictly correct they are the terms used in a surviving order of 31 October 1944 issued to the firm of Henschel und Sohn by the Heeres-Abnahmestelle II(K) which is probably the best evidence we have for this scheme and so they are rendered here verbatim.
2. Zimmerit, the anti-magnetic mine paste applied to German tanks and assault guns from September 1943, was never incorporated into the production of the Hetzer.

vehicle undercoated in either Dunkelgrün or Rotbraun in monochrome photographs (1). In any case the order was countermanded on 23 January 1945, advising manufacturers that existing paint stocks could be used until 30 May 1945, provided that they were not thinned (2). It was anticipated that the new camouflage would be in general use by June 1945. It has been suggested by at least one authoritative source that white, or perhaps Elfenbein RAL 1001, may have been used as a camouflage colour during late 1944 and early 1945 and this is discussed in the Camouflage & Markings section. In September 1944 BMM was able to complete 190 vehicles while and a further thirty were assembled by Skoda and a total of 158 Hetzers were dispatched to units by the HZA. In October, total production fell to 190, partly as a result of two Allied air raids on the Skoda plant in that month, and the HZA shipped 224 Hetzers to operational units. In November a further 317 Hetzers and Bergepanzer 38 recovery vehicles were dispatched by the HZA. During December BMM and Skoda managed to assemble 327 vehicles and a total of 323 Hetzers, including twenty Flammpanzer 38 flamethrower tanks, and sixteen Bergepanzers were shipped to units.

4. January to April 1945. The Final Production Model.

The standard drive sprocket and armoured final drive housing were retained. The 16-bolt roadwheels had probably by now replaced all stocks of the earlier 32-bolt models. A rear idler with just four holes, in both dished and flat versions, was introduced but may have been restricted to Skoda produced vehicles. The towing eyes at both the front and rear were strengthened with a triangular brace. Towards the middle of this production run the hull sides were no longer extended and a simple U-shaped bracket was welded to the hull. The Type V Kugellaffette and late gun mantlet were retained.

Metal rings were welded to the hull front and sides to hold foliage camouflage. The shovel and sledgehammer were removed from the engine compartment and stowed on the right side of the hull below the crowbar. Clips were added to hold them in place. In an effort to simplify production all hinges were welded rather than bolted. The radiator filler hatch on the right side was welded shut and incorporated into the main engine hatch. In April 1945 a number of the vehicles assembled at Skoda were built with a Flammvernichter exhaust muffler fitted in line with the body rather than at 45 degrees. In addition, the right side filler hatch featured a single hinge. The switch to a diesel engine, ordered by the Inspekteur der Panzertruppen on 19 March 1945, was abandoned as it would have disrupted production.

Issued on 3 February 1945, Heeres-Verordnungsblatt Nr.52 clarified the previous orders regarding vehicle camouflage and reiterated that the base coat was to be Dunkelgrün, camouflaged with patches of Rotbraun and Dunkelgelb applied in equal amounts (3).

Notes

1. It has been suggested that stocks of Czech army green paint, a somewhat lighter shade than Dunkelgrün, was used at BMM and Skoda to comply with this order. The evidence for this case seems to rest on a single photograph of a damaged Hetzer where a fourth dark camouflage colour is visible.
2. This order was in fact a confirmation of a memorandum dated 25 December 1944 and headed Anstrich des Heeresgerät, or Painting of Army Equipment. The implication here that manufacturers had been thinning the paint may explain why the contrast between the camouflage colours is quite low in some photographs and why others appear to show a four-colour pattern.
3. Heeres-Verordnungsblatt were orders dealing with unclassified information. Although the word Dunkelgrün is used in the order the base colour is also identified as RAL 6003, the correct code for Olivgrün.

This Hetzer was photographed in Italy in the last weeks of the war and the image provides a good view of the rear hull and engine deck and also shows a mixture of mid, late and final production features which include: A. The metal foliage rings. B. The two-piece commander's hatch. C. The rear periscope guard. D. The right side filler hatch, here open. The Flammvernichter exhaust muffler. F. The left side filler hatch and handle. G. The rear driving light. H. The solid-sided toolbox. I. A non-standard rack. J. The 8-hole rear idler. K. The hull Schürzen do not have the bend on the rear and front sections and the brackets are held in place by rivets. These features and their introduction are explained in the accompanying text.

Photographed in April 1945, this Jagdpanzer 38 Hetzer has the large sheet metal cover over the driver's visor that is indicative of final production vehicles. Note the prominent weld bead across the top of the Kugellaffette. Another view of this tank destroyer is shown in the Camouflage & Markings section on page 25.

The camouflage was to be completed after final assembly and a template or guide was provided. Rather ambiguously, the order emphasised that the use of Dunkelgelb as a base colour was to cease but went on to stress that deliveries to front-line units were not to be delayed due to the lack of suitable colour paint. Interestingly, the dark yellow colour is referred to as 'Dunkelgelb RAL 7028 Ausgabe 1944' suggesting that this shade may have been somehow different to the colour stipulated in April 1943. The exact shade of Dunkelgelb and the use of this colour has perhaps generated more debate than any other aspect of German tank development.

The assembly plants reached their peak production in January, turning out 434 Hetzers and Bergepanzer 38s while the HZA dispatched a total of 420 vehicles with most going to infantry units serving in the East. This was a remarkable feat considering the strategic situation but still short of the 500 vehicles per month envisaged by the HWA at the start of production. This total included thirty-two Hetzers that that had been returned from repair and it is likely that these vehicles featured a number of early parts and modifications.

In February the HZA shipped a total of 351 vehicles although at least fifty of these did not reach their allocated units until the following month and a number were eventually distributed to other formations. The total number included thirty vehicles which had been returned to service from the repair depots which may have all gone to Panzerjäger-Abteilung 744. In addition, forty-three Bergepanzer 38 were dispatched, mostly to units fighting on the Eastern Front, and this is the first time that the records show separate allocations. A total of 490 Hetzers were shipped during March together with twenty-one Bergepanzer 38 recovery vehicles. Although the BMM plant had been extensively damaged in an Allied air raid on 25 March, official records show that seventy vehicles were completed in April. This total may include a number of Bergepanzer recovery vehicles.

The Skoda plant ceased production on 24 April 1945 after assembling forty-seven Hetzers. It is not certain how many of these tank destroyers were accepted by the Heereswaffenamt but in spite of the disruptions to production, a total of 151 Hetzers and twelve Bergepanzer 38 recovery vehicles were shipped, all going to front-line units. This total included ten vehicles taken from the training facility at Milowitz, to where BMM production had been moved. Planned production for May 1945 was uncertain as the manufacturers reported that just fifteen 7.5cm Pak 39 guns were available and these had neither sights nor traversing gear. Permission was granted for the necessary parts to be removed from the Jagdpanzer 38 Starr prototypes and also to mount 7.5cm StuK 40 guns in any finished Hetzer hulls (1). This would have given a possible total of ninety assembled vehicles but it is not known how many, if any, were completed.

Notes

1. The Jagdpanzer 38 Starr was to be the final development of the Hetzer, armed with a gun which lacked a recoil mechanism. Ten examples were built of which nine were converted to standard Hetzers. The 7.5cm StuK 40 gun was being built at Skoda at the time and was in plentiful supply.

At left and below: Partly-completed Jagdpanzer 38 Hetzers photographed the Skoda assembly plant at Pilsen in May 1945. Note the Flammvernichter muffler set at 90 degrees and the 4-hole rear idlers. The latter may have only been in use by Skoda. Note also that the camouflage scheme seems to have been extended to the lower hull behind the wheels which are painted in a very light shade, probably Dunkelgelb RAL 7028. These images were taken from an extensive series in which all the vehicles appear to have been built with the hull extension towing eyes suggesting that this may have been a feature that was retained at Skoda until the end of production.

Below: Also photographed at the Skoda plant at the same time as the vehicles shown above, this Hetzer provides a good view of several late and final production modifications. These include: A. Hull brackets welded to the inside face of the Schürzen. B. A very late, perhaps final, model track. Note the two notches on the forward edge of each link. C. The triangular metal brace welded to the hull extension. D. Metal rings to hold foliage camouflage. E. Large sheet metal cover for driver's visor. F. Kugellaffette V ball mount. Note also the black rectangles painted onto the hull glacis to distract the attention of enemy gunner's from the driver's visor.

Dragon Models Ltd
B1-10/F., 603-609 Castle Peak Rd,
Kong Nam Industrial Building,
Tsuen Wan, N. T., Hong Kong
www.dragon-models.com

Tamiya Inc
Shizuoka City, Japan
www.tamiya.com

Italeri S.p.A.
via Pradazzo 6/b,
40012 Calderara di Reno, Bologna, Italy
www.italeri.com

Revell
Revell USA
www.revell.com
Revell Germany
www.revell.de

Airfix
www.airfix.com

Academy Plastic Models
521-1, Yonghyeon-dong, Uijeongbu-si,
Gyeonggi-do, Korea
Www.academy.co.kr

Hauler
Jan Sobotka,
Moravská 38, 620 00 Brno,
Czech Republic
www.hauler.cz

Sol Model (Sol Resin Factory)
111-45, Wonjuk-ro, Shindun-myun, Icheon-si,
Kyungki-do, Republic of Korea
www.solmodel.com

Voyager
Room 501, No.411 4th Village,
SPC Jinshan District, Shanghai 200540,
P.R.China
www.voyagermodel.com

Aber
ul. Jalowcowa 15, 40-750 Katowice,
Poland
www.aber.net.pl

Friulmodel
H 8142. Urhida, Nefelejcs u. 2., Hungary
www.friulmodel.hu

Modelkasten
Chiyoda-ku Kanda, Nishiki-Cho 1-7, Tokyo,
Japan
www.modelkasten.com
Very difficult to navigate but worthwhile.

Master Club
www.masterclub.ru
It appears that this firm is closely associated
with Armour35, a Russsian mail-order firm.

Voyager
Room 501, No.411 4th Village,
SPC Jinshan District, Shanghai 200540
P.R.China
www.voyagermodel.com

OKB Grigorov
www.shop.okbgrigorov.com

Eduard Model Accessories
Mirova 170, 435 21 Obrnice,
Czech Republic
www.eduard.com

Royal Model
Via E. Montale, 19-95030 Pedara, Italy
www.royalmodel.com

RB Model
Powstancow Wlkp.29B,
64-360 Zbaszyn,
Poland
www.rbmodel.com

Passion Models
Availble from MS Models Japan
4841-10 Nakane, Hitachinaka,
Ibaraki 312-0011, Japan
www.msmodelswebshop.jp

Griffon Model
Suite 501, Bldg 01, 418 Middle Longpan Rd,
Nanjing, P.R. China
www.griffonmodel.com

Modelkasten
Chiyoda-ku Kanda, Nishiki-Cho 1-7, Tokyo,
Japan
www.modelkasten.com
Very difficult to navigate but worthwhile.

Armory Models Group
76014 Ivano-Frankivsk, Ukraine
www.armorymodels.com

In researching this project I consulted a number of works including *Jagdpanzer 38 Hetzer* by Vladimir Francev and Miroslav Bílý, *Jagdpanzer 38 Hetzer* by Hilary Doyle and Tom Jentz and *Jadgpanzer 38 Hetzer Walk Around* from Hans-Heiri Stapfer. Also helpful was *Die Panzer-Kampfwagen 35(t) und 38(t) und ihre Abarten* by Walter J. Spielberger and although somewhat dated this book does contain some very useful period photographs of a number of technical details. The excellent Panzerwrecks series of books include a large number of interesting photographs and the ongoing research, often from readers, is to be commended. I would very much like to acknowledge the work of a number of researchers such as Martin Block, James Blackwell and the late Ron Klages which proved invaluable. I would also like to thank the modellers who very generously allowed me to publish the images of their work and, as always, I am indebted to my friends and long-time associates Karl Berne, Valeri Polokov and J. Howard Parker for their invaluable assistance with the photographs and period insignia.

Below: Photographed in Prague in April or May 1945, this Jagdpanzer 38 has the metal foliage rings on the hull side suggesting it was assembled in early 1945. But the commander's hatch is the earlier type and the filler hatches have bolted hinges which were later simply welded. This mixture of production features is commonly seen. The style of company number and the large wooden stowage box were features common to the Hetzers shown and discussed on page 11 of the main text and page 24 of the Camouflage & Markings section and there could be some connection.